Edwin Lankester

Half-Hours with the Microscope

Being a popular guide to the use of the microscope as a means of amusement and instruction

Edwin Lankester

Half-Hours with the Microscope
Being a popular guide to the use of the microscope as a means of amusement and instruction

ISBN/EAN: 9783337075385

Printed in Europe, USA, Canada, Australia, Japan

Cover: Foto ©berggeist007 / pixelio.de

More available books at **www.hansebooks.com**

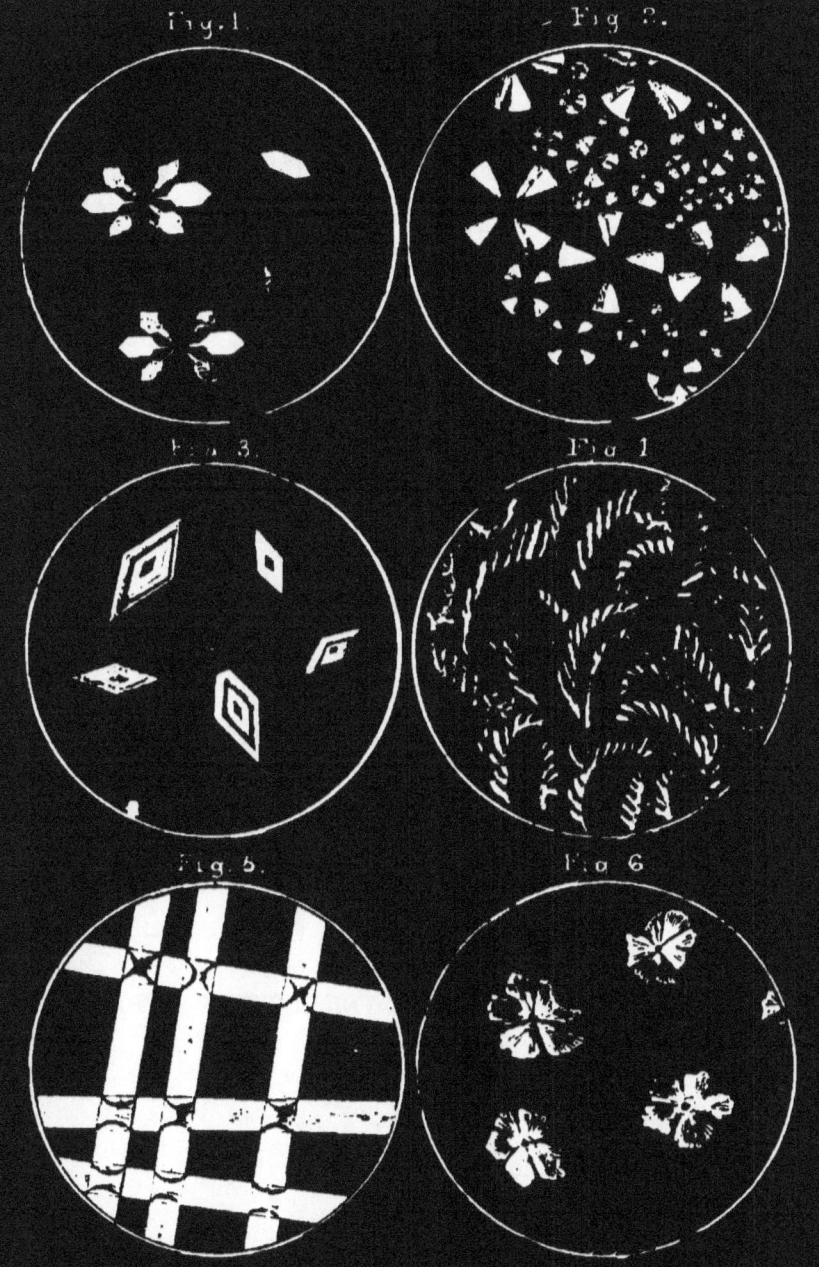

HALF-HOURS

WITH

THE MICROSCOPE;

BEING A POPULAR GUIDE TO THE USE OF THE MICROSCOPE AS A MEANS OF AMUSEMENT AND INSTRUCTION.

BY EDWIN LANKESTER, M.D.

ILLUSTRATED FROM NATURE,

BY

TUFFEN WEST.

A NEW EDITION.
With Chapter on the Polariscope by F. Kitton.

NEW YORK:
G. P. PUTNAM'S SONS,
FOURTH AVENUE AND TWENTY-THIRD STREET.
1874.

CONTENTS.

CHAPTER I.
A HALF-HOUR ON THE STRUCTURE OF THE MICROSCOPE ..... PAGE 1

CHAPTER II.
A HALF-HOUR WITH THE MICROSCOPE IN THE GARDEN 30

CHAPTER III.
A HALF-HOUR WITH THE MICROSCOPE IN THE COUNTRY 47

CHAPTER IV.
A HALF-HOUR WITH THE MICROSCOPE AT THE POND-SIDE 56

CHAPTER V.
A HALF-HOUR WITH THE MISCROSCOPE AT THE SEA-SIDE 67

CHAPTER VI.
A HALF-HOUR WITH THE MICROSCOPE IN-DOORS 78

CHAPTER VII.
A HALF-HOUR WITH POLARIZED LIGHT 98

APPENDIX.
THE PREPARATION AND MOUNTING OF OBJECTS 119

DESCRIPTION OF PLATES.

In the examination of these Plates the observer is requested to remember that they are not all drawn to the same scale. Some objects, adapted for low powers, are only magnified a few times, whilst smaller objects are magnified many hundred times. All objects, of course, vary in apparent size, according to the powers with which they are examined. Descriptions of the objects will be found in the pages indicated.

PLATE I. to face page 1.

FIG.		PAGE
1.	Vegetable cells with nucleus from apple	31
2.	Cellular tissue from pith of elder	31
3.	Stellate cell-tissue from rush	32
4.	Flat tabular cell from surface of tongue	91
5.	Ciliated cell from windpipe of calf	91
6.	Human blood corpuscles	91
7.	Blood corpuscles from fowl	92
8.	Blood corpuscle from frog	92
9.	Blood corpuscle from sole	92
10.	Blood corpuscle from beetle	92
11.	Filament of a species of *Zygnema*, a plant	60

    a. Portion of a filament of the same, the cell-contents becoming changed into zoospores.

    b. Zoospore more highly magnified.

FIG.		PAGE
12. Filament of a species of *Oscillatoria*, a plant	....	60

 a. Portion more highly magnified.

13. *Pandorina Morum*, a plant.................... 60
14. *Volvox Globator*, a plant...................... 60
15. *Euglena viridis*, a plant, showing various forms which it assumes 61
16. *Amœba*, an infusory animalcule 69

 a, b, c, show the various forms which this animalcule assumes........................

17. *Actinophrys Sol*, the sun animalcule 62-69
18. *Difflugia*, an infusory animalcule................ 63
19. *Arcella*, an infusory animalcule 63
20. *Lagena*, a species of Foraminifer................ 69
21. *Polystomella crispa*, a species of Foraminifer...... 69
22. *Globigerina*, a species of Foraminifer 69
23. *Rosalina*, from chalk, a Foraminifer 69
24. Living *Rosalina*, a Foraminifer 69
25. *Textilaria*, a species of Foraminifer.............. 69

PLATE II. to face page 32.

26. *Ulva* in different stages of development........... 61

 a. Cells in single series.
 b. Commencement of lateral extension.
 c. Portion expanded.

27 *Cosmarium*, a species of Desmid undergoing self-division.
28. *Euastrum*, a species of Desmid................... 57
29. *Closterium*, a species of Desmid 57

 a. Undergoing self-division.

30. *Desmidium*, a species of Desmid 57

DESCRIPTION OF PLATES.

FIG.		PAGE
31.	*Pediastrum*, a species of Desmid	57
32.	*Scencdesmus*, a species of Desmid	57
33.	*Surirella nobilis*, a species of Diatom	59
34.	*Pinnularia viridis*, a species of Diatom	59
35, a.	*Navicula*, a species of Diatom undergoing self-division	
b.	Front view of the same.	
36.	*Melosira varians*, a species of Diatom	59
37.	*Melosira nummuloides* undergoing self-division	59
38.	*Coscinodiscus eccentricus*, a species of Diatom	58
39.	*Paramecium Aurelia*, an infusory animalcule	64
40.	*Vorticella nebulifera*, an infusory animalcule	63
41.	*Rotifer vulgaris*, a wheel animalcule	65
42.	Stomates on a portion of cuticle of hyacinth leaf	32
43.	Sinuous walled cells and stomates from under surface of leaf of water-cress	32
44.	Cuticle of wheat straw with stomates	33
45.	Cuticle from petal of geranium (*Pelargonium*)	33
46.	Cuticle from leaf of a species of aloe	33
47.	Spiral vessel from leaf-stalk of garden rhubarb	35
48.	Ditto unrolled	35
49.	Annular vessel from wheat root	35
50.	Dichotomous spiral vessels	35
51.	Dotted duct from common radish	35
52.	Scalariform tissue from fern root	35
53.	Woody fibre from elder	35

PLATE III. to face page 40.

FIG.		PAGE
54.	"Glandular" woody tissue	34
55.	Transverse section of glandular woody tissue	34
56.	Transverse section of oak	34
57.	Long section of oak	34
58.	Oblique section of oak	34
59.	Section of cork	35
60.	Transverse section of coal	36
61.	Longitudinal section of coal	36
62.	Wheat starch	37
63.	Oat starch	37
64.	Potato starch	37
65.	Tous-les-mois starch	37
66.	Indian corn starch	38
67.	Sago starch	37
68.	Tapioca starch	37
69.	Acicular raphides from garden hyacinth	38
70.	Bundle of ditto from leaf of aloe contained in a cell	38
71.	Compound raphides from stalk of garden rhubarb	39
72.	Tabular prismatic raphides from outer coat of onion	39
73.	Circular crystalline mass from a cactus	39
74.	Simple vegetable hair from leaf of a common grass	40
75.	Rudimentary hair from flower of pansy	40
76.	Simple club-shaped hair	40
77.	Club-shaped hair from leaf of dock	41
78.	Hair from throat of pansy	40
79, a.	Hair formed of two cells from flower of white dead-nettle	41
79, b.	Many-jointed tapering hair with nuclei from common groundsel	41

DESCRIPTION OF PLATES.

FIG.		PAGE
80.	Beaded hair of sow-thistle	41
81.	Glandular hair from leaf of common tobacco	41
82.	Hair from leaf of garden chrysanthemum	41
83.	Rosette-shaped glandular hair from flower of verbena	41
84.	Stellate hairs from the hollyhock (*Althæa rosea*)	41
85, a.	Stellate hair from leaf of lavender	41
85, b.	Hair from leaf of garden verbena, with warty surface	41
86.	Hair from leaf of white poplar (*Populus alba*)	41
87.	Base of a hair on a mass of cellular tissue	41
88, a.	A sting from common nettle	42
88, b.	Portion of a leaf of Valisneria	42

PLATE IV. to face page 48.

89.	*Palmella cruenta*—gory dew	48
90.	Yeast plant	48
91.	Portions of vinegar plant	48
92.	So-called cholera fungus obtained from the air	48
93.	Red rust of wheat	49
94.	*Puccinia graminis*—mildew	49
95.	*Penicillium glaucum*—common mould	49
96.	*Botrytis* from mouldy grape	49
97.	Fungus from mouldy bread (*Mucor Mucedo*)	49
98.	Fungus from human ear	49
99.	Fungus from leaf of bramble (*Phragmidium bulbosum*)	49
100.	Vine blight (*Oidium Tuckeri*)	50
101.	Potato blight (*Botrytis infestans*)	50
102, a.	Pea-blight (*Erysiphe Pisi*)	50
b.	*Asci* and *sporidia* of pea blight	50
103.	Fungus from a decayed Spanish nut	50

FIG.		PAGE
104.	Curious fungus from oil casks	50
105.	Fungus of common ringworm (*Achorion Schönlenii*)	50
106.	Fungus on stem of duckweed	50
	a. Another within the cells.	
107, *a.*	Branched cells from stem of mushroom	51
	b. Branched cells from rootlets of mushroom	51
	c. Reproductive bodies borne in fours on the gills of mushrooms	51
108.	Section through a brilliant orange-coloured peziza	52
109.	Section through the common yellow lichen of trees and walls	52
110.	Leaf of *Sphagnum*—bog moss	52
111.	Sea weed—*Polysiphonia fastigiata*	67
	a. Fruit-bearing organs.	
	b. Spore.	
	c. Portion of Bisporo; and	
	d. „ Tetraspore.	
	e. Antheridia.	
112.	Reproductive organs of a moss, a species of *Tortula*	52
	a. The calyptra.	
	b. The operculum.	
	c. The peristome.	
	d. The teeth.	
	e. The spores.	
113.	Fructification on back of frond of male fern	53
114.	Fructification on back of froud of common brakes	53
115.	Capsules of *Scolopendrium*—hartstongue. The sporules seen escaping	53
	a. One of the latter more magnified	53
116.	Fructification of *Equisetum*—horsetail	55
	a. Shield-like disk of ditto, separated, surrounded by thecæ	55

| FIG. | PAGE |

116, *b.* Spore, much magnified, with elastic filaments coiled closely round 55
    c. Spore expanded 55
117, *a.* Fructification of *Lycopodium*—club moss...... 54
    b. Sporules 54
    c. Sporules more highly magnified.

PLATE V. to face page 56.

118. Delicate spiral cells from anthers of furze........ 44
119. Large well-developed spiral cells from anthers of hyacinth, with minute raphides in intercellular spaces.. 44
120. Irregular deposit in cells of anthers of white dead-nettle.. 44
121. Annular ducts from anthers of narcissus 44
122. Stellate cells from anthers of crown imperial 44
123. Ovate pollen cells............................. 44
124. Triangular pollen cells from hazel 44
125. Pollen cells of heath 44
126. Pollen cells of dandelion...................... 44
127. Pollen cells of passion flower 45
128. Pollen cells of mallow........................ 45
129. Red poppy seed............................... 46
130. Black mustard seed 46
131. Seed with deep and curved furrows 46
132. Great snapdragon seed 46
133. Chickweed seed............................... 46
134. Umbelliferous seed or fruit 46
135. *Zygnema*, conjugating 60
136. *Closterium*, conjugating 57

FIG.		PAGE
137.	*Cosmarium*, conjugating	57
138.	*Epithemia gibba*, conjugating	43
139.	*Melosira nummuloides*, conjugating	59
140.	Transverse section of common sponge	63
141.	Transverse section of common British sponge	63
	a. Spicules of the same more magnified.	
	Calcareous spicules of *Grantia ciliata*.	68
142.	Pin-like spiculum from *Cliona*, a boring sponge	69
143.	Spiculum from *Spongilla*, a fresh-water sponge	69
144.	Spiculum from unknown sponge	69
145.	Spiculum from *Tethea*	69
146.	Common *Hydra*	70
	a. Stinging organ from common *Hydra*	
147.	A species of *Sertularia*, a zoophyte	71
148.	*Campanularia integra*, a zoophyte	71
149.	"Cup" of *Campanularia volubilis*, a zoophyte	71
150.	Spicula of *Gorgonia verrucosa*	71
151.	Transverse section from base of spine of *Echinus neglectus*	72
152.	Calcareous rosette from sucker of *Echinus*	72
153.	Pedicellaria from *Echinus*	72
154.	Pedicellaria from star-fish	72

PLATE VI. to face page 72.

155.	*Lepralia*, a polyzoon	72
156.	*Bowerbankia densa*, a polyzoon	73
157.	Tobacco-pipe, or bird's-head processes of *Notamia*	73
	a. Bird's-head process.	
158.	*Bugula avicularia*	73
159.	Bird's-head process of *Bugula Murrayana*	73

FIG.		PAGE
160.	*Scrupularia scruposa*, with bird's-head processes (*avicularia*) and sweeping bristles (*vibracula*)..	73
161.	Snake-headed zoophyte—*Anguinaria*	73
162.	*Flustra foliacea*—sea mat	72
163.	*Plumatella repens*, a fresh-water polyzoon........	74
164.	Egg of *Cristatella Mucedo*, a fresh-water polyzoon	74
165.	Transverse section of shell of *Pinna*, showing prismatic shell structure	74
166.	Longitudinal section of shell of *Pinna*	75
167.	Transverse section from oyster-shell	75
168.	Section of shell of *Anomia*, with tubular borings..	75
169.	Section of mother of pearl	75
170.	Prawn-shell viewed as a transparent object	75
171.	Teeth of whelk................................	76
172.	Teeth of limpet...............................	76
173.	Teeth of periwinkle	76
174.	Teeth of *Limneus*	76
175.	Scale of sturgeon—*ganoid*.....................	76
176.	Prickle from back of skate—*placoid*	76
177.	Borings by a minute parasite in a fossil fish-scale..	77
178.	Scale of sole—*ctenoid*	77
179.	Scale of whiting—*cycloid*......................	77
	a. Calcareous particles, magnified.	
180.	Scale of sprat—*cycloid*	77
181.	Section of egg-shell	90
182.	From soft egg.	
183.	Section of egg-shell of emu....................	90

PLATE VII. to face page 80.

184.	Human hair	78
	a. Transverse section of human hair.	

FIG.		PAGE
185,	a. Small mouse-hair	79
	b. Larger mouse-hair.	
	c. Plain mouse-hair.	
	d. Minute hair from ear of mouse.	
186.	Hair of long-eared bat	80
187.	Transverse section of hair of peccary	80
188.	Pith-like hair of musk-deer	80
189.	Hair from tiger caterpillar	80
190,	a. Branched hairs from leg of garden spider (*Epeira diadema*)	81
	b. Spine, with spiral flutings, from the same.	
	c. Small brush-like hairs from an Australian spider.	
191.	Hair from flabellum of crab	81
192.	Portion of four of the barbs of a goose-quill	81
193.	Portion of the same more magnified	81
194.	Swan's-down	81
195.	Head and mouth of a flea	82
196.	Head and mouth of a bug	82
197.	Mandible of humble bee	84
198.	Head and mouth of louse	83
199.	Head and mouth of gnat	83
200.	Extremities of barbs of the sting of common bee	84
201.	Head of honey bee	83
	a. Piece of the tongue more magnified.	
202.	Mouth of blow-fly	84
203.	Head and mouth of butterfly	84
204.	One of the fangs of a spider, showing the poison-bag and duct	85
205.	Foot of Empis, a species of fly	87
206.	Foot of bee	87
207.	Foot of spider	87

FIG.		PAGE
208.	Head of common spider, showing eight simple eyes.	
	a. Cornea of one of these more magnified......	85
209.	Skin of garden spider	85
210.	Portion of compound eye of fly	83
211.	Portions of the two wings of bee in flight........	88
	a. Nervule of wing.	
212.	Spiracle of fly	86
213.	Spiracle of *Dytiscus*	86
214.	Threads of garden spider (*Epeira diadema*)	86
	Simple thread of the same.	
	Thread of a concentric circle with viscous dots.	

PLATE VIII. to face page 88.

215.	Fore leg of *Gyrinus natator*, whirligig beetle	87
216.	Middle leg of the same	87
217.	Hind leg of same	87
218.	Fore leg of male *Dytiscus*	87
219.	Middle leg of the same.	
220, a.	Gizzard of cockroach	89
b.	Ditto, cut open.	
221, a.	Gizzard of cricket	89
b.	Ditto, cut open.	
222.	Trachea from caterpillar	86
223.	Proleg of caterpillar of common garden white butterfly, with the membranes in which the hooks are seated, expanded as in action	83
224.	Part of leg of cockroach	
225.	Battledore scale from blue argus butterfly	88
226.	Scale of ordinary shape from same.............	88
227.	Scale from meadow-brown butterfly	88
228.	Scale of gnat................................	88

FIG.		PAGE
229.	Scale reduced to a hair from clothes-moth........	85
230.	Hair-like scale from clothes-moth, with three prongs.................................	89
231.	Cartilage from mouse's ear....................	91
232.	Transverse section of human bone..............	90
233.	Striped muscular fibre from meat..............	92
234.	a. Liber fibre of flax, natural state .... b. Ditto, broken across at short intervals	79
235.	Wool from flannel.............................	79
236.	Silk..	79
237.	Cotton hair...................................	79
238.	Crystal of honey	39
239.	Thick crystal of ordinary sugar—same angles....	39
240.	Crystals of sugar from adulterated honey	40
241.	Cuticle from berry of holly	33
242.	Transverse section of whalebone	90
243.	Transverse section of plum-stone	31
244.	Transverse section of testa of seed of Guelder rose	13
245.	Fruit of groundsel—opaque	42
246.	One hair of pappus of dandelion................	42
247.	Cottony hair of burdock	42
248.	Portion of pappus of goats-beard	42
249.	Wood of young shoots of vine, the cells containing starch	33
250.	Spiral fibres from testa of wild sage seed	35

PLATE IX.   Frontispiece to face title-page.

1. Iodo-sulphate of Quinine
2. Salicine
3. Aspartic Acid
4. Sulphate of Copper in Gelatine
5. Grey Hair (human)
6. Scales of Hyppophæ rhamnoides

PLATE I

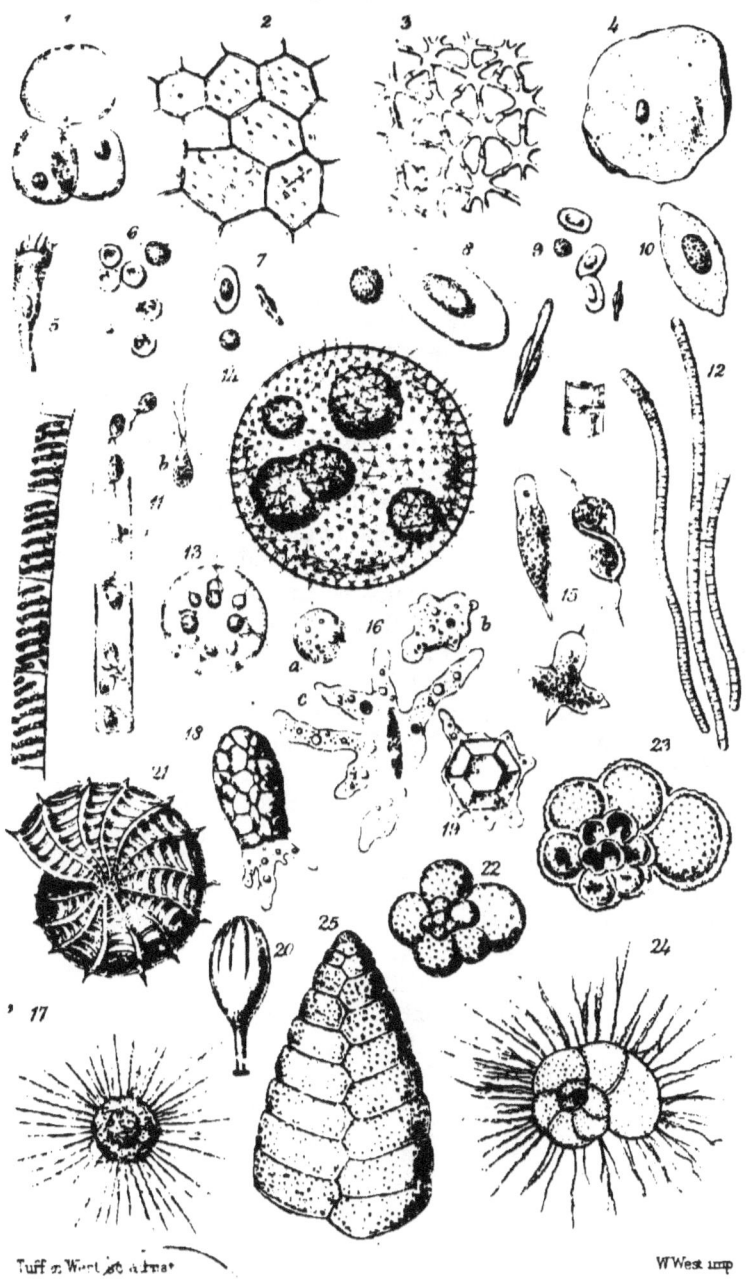

London: Robert Hardwicke, 1860

HALF-HOURS WITH THE MICROSCOPE.

CHAPTER I.

A HALF-HOUR ON THE STRUCTURE OF THE MICROSCOPE.

The Microscope is often regarded merely as a toy, capable of affording only a certain amount of amusement. However much this might have been the case when its manufacture was less perfectly understood, it is now an instrument of so much importance that scarcely any other can vie with it in the interest we attach to the discoveries made by its aid. By its means man increases the power of his vision, so that he thus gains a greater knowledge of the nature of all objects by which he is surrounded. What eyes would be to the man who is born blind, the Microscope is to the man who sees only with his naked eye. It opens a new world to him, and thousands of objects whose form and shape, and even existence, he could only imagine, can now be observed with accuracy.

Nor is this increase of knowledge without great advantages. Take for instance the study of plants and animals. Both are endowed with what we call life: they grow and perform certain living functions; but as to the mode of their growth, and the way in which their functions were performed, little or nothing was known till the Microscope revealed their minute structure, and showed how their various parts were related to each other. The

Microscope has thus become a necessary instrument in the hands of the botanist, the physiologist, the zoologist, the anatomist, and the geologist.

Let us, then, endeavour to understand how it is this little instrument has been of such great service in helping on the advancement of science. Its use depends entirely on its assisting the human eye to see—to see more with its aid than it could possibly do without it. This it does by enabling the eye to be brought more closely in contact with an object than it otherwise could be.

Just in proportion as we bring our eyes close to objects, do we see more of them. Thus, if we look at a printed bill from the opposite side of a street, we can see the larger letters only; but if we go nearer we see the smaller letters, till at last we get to a point when we can see no more by getting closer. Now suppose there were letters printed on the bill so small that we could not see them with the naked eye, yet, by the aid of a lens—a piece of convex glass—we could bring our eyes nearer to the letters, and see them distinctly. It would depend entirely on the form of the lens, as to how close we could bring our eyes to the print, and see; but this great fact will be observed, that the nearer we can get our eyes to the print, the more we shall see. The most important part of a Microscope, then, consists of a lens, by means of which the eye can be brought nearer to any object, and is thus enabled to see more of it. Magnifying-glasses and Simple Microscopes consist mainly of this one element. In order, however, to enable the eye to get as close as possible to an object, it becomes convenient to use more than one lens in a glass through which we look. These lenses, for the sake of convenience, are fixed in a brass frame, and attached to the Simple Microscope; when there are two lenses they are called doublets, and when three they are termed triplets. The magnifying-glasses which are made to be held in the hand, frequently have two or three lenses, by which their power may be increased or decreased. Such instruments as these were the first which were employed by microscopic observers; and it is a proof of the essential nature of this part of the Microscope, that many of the greatest discoveries have been made with the Simple Microscope.

The nearer the glass or lens is brought to an object, so as to enable the eye to see, the more of its details will be observed. So that when we use a glass which enables us to see within one inch of an object, we see much more than if we could bring it within only an inch and a half or two inches. So on, till we come to distances so small as the eighth, sixteenth, or even twentieth of an inch.

Although a great deal may be seen by a common hand-glass, such as may be purchased at an optician's for a few shillings, yet the hand is unsteady; and if these glasses were made with a very short focus, it would be almost impossible to use them. Besides, it is very desirable, in examining objects, to have both hands free. On these accounts the glasses, which in such an arrangement are called *object-glasses* (see fig. 3), are attached to a *stand*, and placed in an arm, which moves up and down with rackwork. In this way, the distance of the object from the glass can be regulated with great nicety. Underneath the glass, and attached to the same stand, is a little plate or framework, to hold objects, which are placed on a *slide* of glass. This is called *the stage*. (Fig. 1, *G*.) Sometimes rack-work is added to this stage, by which the objects can be moved upon it backwards and forwards, without being moved by the hand. Such an arrangement as this is called a *Simple Microscope*. Of course many other things may be added to it, to make it more convenient for observation; but these are its essential parts.

But, although the Simple Microscope embraces the essential conditions of all Microscopes, and has, in the hands of competent observers, done so much for science, it is, nevertheless, going out of fashion, and giving way to the *Compound Microscope*. (Fig. 1, p. 5.) This instrument, as might be inferred from its name, is much more complicated than the Simple Microscope, but it is now constructed with so much accuracy, that it can be used with as great certainty and ease as the Simple Microscope itself. In order to understand the mechanism of the Compound Microscope we must first of all study the principles on which it is constructed. If we take a common convex lens and place any small object on one side of it, so as to be in its focus, and then place on the other side a sheet of white paper, we shall find at a certain point that an enlarged picture of the object will be produced on the paper; and this is the way in which pictures are formed by the camera of which the photographic artist avails himself for his portraits and sun-pictures. Now if we look at this picture with another lens of the same character but of somewhat less magnifying power, we shall obtain a second picture larger than the first, and this is the principle involved in the Compound Microscope. The superiority of this instrument over the Simple Microscope consists in an increase of magnifying power. There is, however, a limit to the utility of this magnifying power; for when objects are greatly magnified they become indistinct. This is seen in the Oxyhydrogen and Solar Microscopes, where the images are thrown, by means of highly magnifying lenses, on a white sheet; and, although made enormously large, their details are much less clear than when looked at by a lens magnifying much less. Another advantage of the

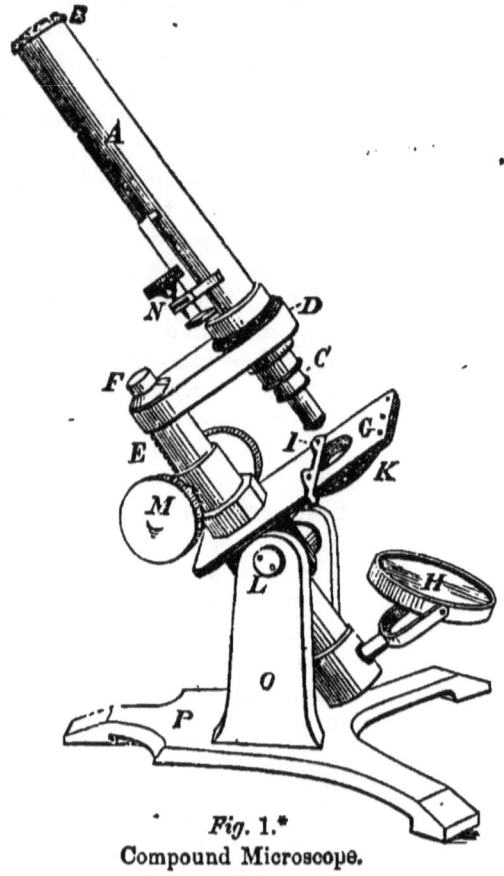

*Fig. 1.**
Compound Microscope.

Compound Microscope is the distance at which the eye is placed from the object, and the facility with

* In this little work we have purposely abstained from mentioning either the names or the Microscopes of our principal makers, lest we should thereby seem to give a

which the hands may be used for all purposes of manipulation.

A brief description, aided by the accompanying illustration, will, it is hoped, suffice to make the beginner acquainted with the various parts of this important instrument.

We have already mentioned that when powerful lenses are used in the examination of small objects the hand is not sufficiently steady to give a firm support to the lens employed, and this is equally true of the hand that holds the object. It is also essentially requisite to have both hands free, for the purpose of manipulation. Hence it becomes necessary to devise some mechanical means for the support of both the lens and the object. How these wants have been supplied by the enterprising skill and ingenuity of our opticians will be best seen as we describe the various parts of which the Compound Microscope consists.

The most important part of the instrument is undoubtedly that which carries the various lenses or magnifying powers. These are contained in the interior of the tube or body, A, which is usually constructed of brass, and from 8 to 10 inches in length. At the upper end of the tube is the eye-piece, B, so named from its proximity to the eye of the observer. It consists of two plano-convex lenses, set in a short piece of tubing, with their flat surfaces turned towards the eye, and at a distance from each other of half their united focal lengths. The first of these lenses is the eye-glass, while that nearest the objective is termed the field lens. The use of the latter is to alter the course

preference to any. The general excellence of these instruments is so well known and the names of their makers are so universal that the student will find no difficulty in providing himself with an efficient instrument at a moderate cost.

of the light's rays in their passage to the eye, in such manner as to bring the image formed by the object-glass into a condition to be seen by the eye-glass. A stop also is placed between the two lenses in such a position that all the outer rays, which produce the greatest amount of distortion, arising from spherical and chromatic aberration, are cut off. The short tube carrying the lenses (fig. 2) slides freely, but without looseness, into the upper end of the compound body, *A*, an arrangement which affords a ready and convenient method for changing the eye-piece.

Compound Microscopes are generally fitted up with two eye-pieces, the one deep and the other shallow. The last has its lenses close together, and magnifies the most, whilst the other has them farther apart, and magnifies less. In the use of these eye-pieces, it should never be forgotten that the one which magnifies least is generally the most trustworthy.

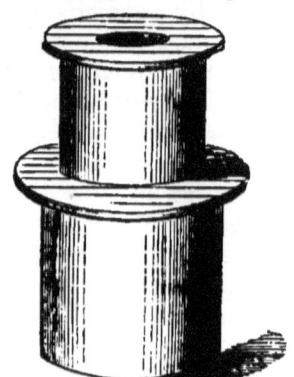

Fig. 2. Eye-piece.

At the opposite end of the tube *A* is the object-glass *C*. The use of this lens is to collect and bring to a point the rays of light that proceed from any object placed in its focus. At this point an enlarged image of the object will be formed in the focus of the eye-glass. We have only to look through the latter at the picture thus formed in order to obtain a second image larger than the first. And this is the way in which minute objects are made to appear so much larger than when seen by the unassisted eye. It will at once be seen how much of the utility of a Microscope depends on good object-glasses. Where they are faulty, the image they form is also faulty; and when these faults in the first image are multiplied by the power of the eye-piece, they become—like the faults of our friends when viewed through a similar medium—of great magnitude.

A good object-glass may be known by its giving a clear and well-defined view of any object we may wish to examine; while a bad lens may be equally well known by the absence of these qualities. In short, a badly constructed objective is more apt to mislead than to guide the student, by the fictitious appearances it creates—appearances that may be erroneously taken for realities, which have no existence in the object itself. The object-glasses of our best opticians consist of several lenses arranged in pairs, set in a small brass tube. A screw at one end serves to attach them to the lower extremity of the compound body, A. (Fig. 3.) The body of the Microscope is supported by a stout metal arm, D, into the free end of which it screws. The opposite end of the arm is secured to the stem, E, by a screw, on which it moves as on a pivot. By this means the tube of the Microscope can be turned away from the stage—an arrangement that gives this form of Microscope an advantage over those that are not so constructed. To the stem, E, which works up and down a hollow pillar by rack-work and pinion, is attached the stage, G. This, in its simplest form, consists of a thin flat plate of brass, for holding objects undergoing ex-

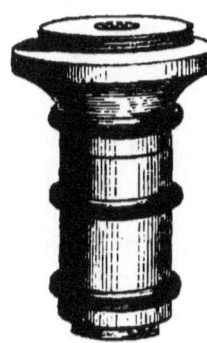

Fig. 3. Object-Glass.

amination. In the centre is a circular opening, for the passage of the light reflected upward by the mirror, H. There is also a sliding ledge, I; against this the glass slide, on which the object is mounted, rests, when the Microscope is inclined from the perpendicular.

In a stage of this kind the various parts of an object can only be brought under the eye by shifting the slide with the fingers. But in more expensive instruments the stage is usually constructed of one or two sliding plates, to which motion is given by rackwork and pinion; the whole being brought under the hand of the operator by two milled heads, a mechanical arrangement which enables him to move with ease and certainty the object he may wish to investigate.

Underneath the stage is the diaphragm, K, a contrivance for limiting the amount of light supplied by the mirror, H. It consists of a thin, circular, flat plate of metal, turning on a pivot, and perforated with three or four circular holes of varying diameter (fig. 4), the largest only being equal

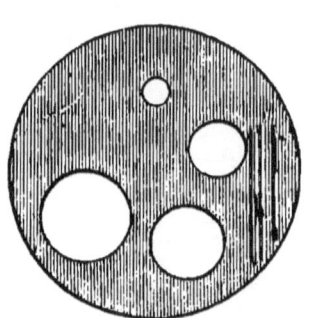

Fig. 4. Diaphragm.

to the aperture in the stage. By turning the plate round, a succession of smaller openings is brought into the centre of the stage, and in one position of the diaphragm the light is totally excluded. By this small but useful contrivance the Microscopist can adjust the illumination of the mirror to suit the character of the object he may be investigating. In some Microscopes the diaphragm is a fixture, but in the better class of instruments it is simply attached to the under part of the stage by a bayonet catch, or by a sliding plate of metal (fig. 5), and can be readily removed therefrom when it is desirable to employ other methods of illumination.

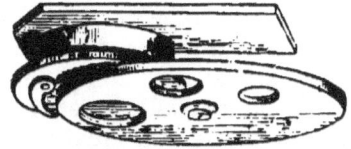

Fig. 5. Diaphragm.

In working with the Microscope it is necessary to adopt some artificial means for ensuring a larger supply of light than can be obtained from the natural diffused light of day, or from a lamp or candle. For this purpose the Microscope is furnished with a double mirror, *H*, having two reflecting surfaces, the one plane and the other convex. The latter is the one usually employed in the illumination of transparent objects; the rays of light which are reflected from its concave surface are made to converge, and thus pass through the object in a condensed form to the eye. The plane mirror is used generally in conjunction with an achromatic condenser, when parallel rays only are required. The whole apparatus is attached to that portion of the hollow pillar continued beneath the stage, in such a manner that it can be moved freely up and down the stem that supports it. This motion enables the Microscopist to regulate the intensity of his light by increasing or decreasing the distance between the mirror and the stage; while the peculiar way in which the mirror itself is suspended on two points of a crescent-shaped arm, turning on a pivot, gives an almost universal motion to the reflecting surfaces. The observer by this means can secure any degree of oblique illumination he may require for the elucidation of the structure undergoing examination.

We next come to the stand, which, though the most mechanical, is at the same time a very important part of the Compound Microscope. On the solidity and steadiness of this portion of the instrument depends in a great measure its utility. The form generally adhered to is that represented in our diagram (fig. 1, p. 5.) It consists of a tripod base, P, from which rise two flat upright pillars, O. Between these, on the two hinge-joints shown at L, is suspended the whole of the apparatus already described: namely, the body carrying the lenses, the arm to which it is attached, the stage, and the mirror underneath it. By this contrivance the Microscope can be inclined at any angle between a vertical and horizontal position—an advantage which can be duly appreciated by those who work with the instrument for two or three hours at a time. Close to the points of suspension are the milled heads, M; these are connected with a pinion working in a rack cut in the stem, E. By turning the milled heads the tube is made to approach or recede from the stage until the proper focus of the object-glass is found. This is termed the *coarse adjustment*, and is generally used for low powers, where delicate focussing is not required. But when high magnifying powers are used, that require a far greater degree of precision, we have recourse to the fine adjustment, N, which consists of a screw acting on the end of a lever. The head of the screw by which motion is communicated to the object-glass is divided into ten equal parts, and when caused to rotate through any of its divisions slightly raises or depresses the tube, carrying the objective with it. As the screw itself contains just 150 threads to an inch one revolution of its head will cause an alteration of the 150th of an inch in the distance of the lens from the object. When moved through only one of its divisions we obtain a result equal to the 1500th of an inch, and by causing it to rotate through half a division we secure a movement not exceeding the 3000th part of an inch in extent. Such nicety in the adjustment of the optical part of the Microscope may seem to the beginner unnecessary, but when he comes to work with high powers he will find that he needs the most delicate mechanical contrivances to enable him to secure the proper focus of a sensitive object-glass.

But this is not the only use to which we can put the fine adjustment. The same process that serves to regulate the focus of a lens will also enable us to measure pretty accurately the thickness of an object or any of the small prominences or depressions found in its structure. By observing the number of divisions through which the head of the screw is made to pass while changing the focus of the object-glass from the bottom to the top of any small cavity or prominence we get a tolerable notion of its depth or height, &c. Connected with this apparatus is a special contrivance for protecting the object-glass to some extent from injury. It will sometimes happen, even with the most careful, when using high powers, that the lens is brought down with some force in contact with the glass cover that protects the object. This risk is not unfrequently incurred by admitting to one's study incautious friends, whose confidence is only equalled by their ignorance; who although they may have never seen a Microscope before, will proceed to turn it up and down with a force sufficient to crack the lens. Such friends would have sufficient confidence in themselves to take the command of a man-of-war, even though it were the first time in their lives they had been on board a ship. Strict injunctions must be laid on all such not to approach the table until the instrument is quite ready for them to take a peep, coupled with a polite request that while doing so they will keep their hands behind them. A provision has been made which to some extent provides for such an emergency. The object-glass itself is screwed into a short tube, that fits accurately the lower end of the compound body and slides freely within it, being kept down in its place by a spiral spring, which presses upon it from behind. On the application of a slight force or resistance to the object-glass the spring tube immediately yields, within certain limits, to the pressure, carrying with it the lens, which is thus often saved from destruction. Object-glasses of various degrees of magnifying power and excellence of workmanship are supplied with the Microscope, and may be purchased separately, according to the wants and resources of the student. It will be found that for all ordinary purposes the 1-inch and ¼-inch objectives are the most useful powers. A substitute for the intermediate powers may be obtained by pulling out the draw-tube and using the higher eye-pieces. This method, though not so satisfactory in its results as the use of separate object-glasses, may be resorted to where a series of objectives are not within the reach of the observer.

THE BINOCULAR MICROSCOPE.

Since the invention of the Stereoscope attempts have been made to apply the Binocular principle in the construction of the Compound Microscope. After some failures this desideratum has been successfully achieved by Mr. F. H. Wenham, a gentleman well known to microscopists by the fertility of his resources and the ingenuity of his inventions in connection with the Microscope. It is to him that we are indebted for a Microscope that enables us to see objects in a natural manner, namely, with both eyes at once. Hitherto the ordinary single-tubed Microscope reduced the observer to the condition of a Cyclops. Although gifted with a pair of eyes he found it impossible to avail himself of this plurality of organs. He was condemned by the very nature of his Microscope to peer perpetually with a single eye through its solitary tube; but thanks to Mr. Wenham all this is changed. We have now the satisfaction of using a double-tubed Microscope that not only gives employment to both eyes at once, but presents us with effects unknown and unattainable by the ordinary instrument. We no longer gaze at a flat surface, but a stereoscopic image stands out before us with a boldness and solidity perfectly marvellous to those who have only been accustomed to the ordinary single-tubed Microscope.

"No one," says a writer in 'The Popular Science Review,' "can fail to be struck with the beautiful appearance of objects viewed under the Binocular Microscope. Its chief application is to such objects as require low powers, and can be seen by reflected light, when the wonderful relief and solidity of the bodies under observation astonish and delight even the adept. Foraminifera, always beautiful, have their beauties increased tenfold; vegetable structures, pollen, and a thousand other things, are seen in their true lights, and even diatoms, we may predict, will receive elucidation, as to the vexed questions of the convexity or concavity of their infinitely minute markings. The importance of the Binocular principle is especially apparent when applied to anatomical investigation. Prepared Microscopic injections exhibit under the ordinary Microscope a mass of interlacing vessels, whose relation, being all on the same plane, it is not easy to make out with any degree of satisfaction. But placed under the Binocular they at once assume their relative position. Instead of a flat band of vessels, we now see layer above layer of tissue; deeper vessels anastomosing with those more superficial; the larger vessels sending branches, some forward and some backward, and the whole injection assumes its *natural* appearance, instead of being only like a *picture*."

Fortunately for the possessors of the ordinary Microscope, the Binocular arrangement can be readily adapted to this instrument at a cost of a few pounds. The additional tube and prism does not interfere with the use of the instrument as a monocular, the withdrawal of the prism instantly converts it into that form of instrument: this is necessary when high powers are used.

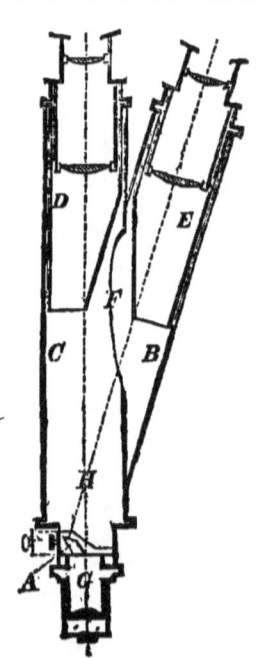

Fig. 6. Section of Binocular Microscope.

The accompanying diagram (fig. 6) — a section of the Binocular — will give the reader a correct notion of the mechanism of the instrument. Let C represent the body of the ordinary Microscope and B the secondary tube attached to the side of the former, which it will be seen has a portion of its surface cut away at the point of junction, F, as a means of communication between them. The eye-pieces and draw-tubes are seen at D and E. The object-glass G is attached to the ordinary tube C in the usual way. Just above it is the small prism, A, mounted in a brass box, and so constructed as to slide into an opening in the tube at the back of the object-glass. By this arrangement it will be found that while one half of the light passes up the tube unobstructed the other half must first pass through the prism, where, after undergoing two reflections (fig. 7), it escapes in the direction of the additional tube B. The dotted lines in the diagram show the direction the light takes in its passage to the eyes. At H the rays are seen to cross each other. Those from the left side of the object-glass traverse the right tube, while those from the right side of the lens are projected up the left tube.

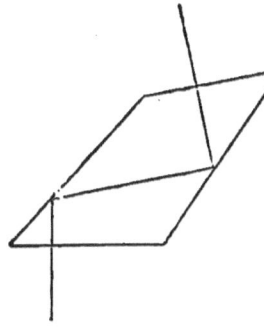

Fig. 7.
Double-reflecting Prism.

In using the Binocular it must be remembered that the eyes of different individuals vary in their distance from each other. It will thus be seen that some contrivance is necessary to enable us to increase or decrease the distance between the eye-pieces to suit the requirements of all. This is accomplished by the two draw-tubes, D and E, which carry the eye-pieces. When drawn out, the latter are made to diverge, and when pushed in they converge. In this way any intermediate distance can be obtained to suit every kind of vision.

Where a Binocular Microscope is in daily use it will sometimes be necessary to withdraw the prism from the tube, to cleanse it from dust and other impurities gradually contracted by use. Whenever this may be necessary great care should be taken to employ no substance likely to scratch its highly polished surfaces; for on these being preserved intact in a great measure depends the efficiency of the instrument. We know of nothing better adapted for removing impurities than a clean silk or cambric handkerchief, which, when not in use, should be kept in a closely-fitting drawer, to protect it from dust.

There seems to be little doubt that this lately improved form of the Compound Microscope will eventually supersede all others. This opinion also seems to be entertained by the inventor himself, whose words we quote :—

"The numerous Microscopes that have been altered into Binoculars, in accordance with my last principle, and also the large quantity still in the course of manufacture, will, I think, justify me in making the assertion, without presumption, that henceforth no first-class Microscope will be considered complete unless adapted with the Binocular arrangement."

The Compound Microscope is now, undoubtedly, one of the most perfect instruments invented and used by man. In the case of all other instruments, the materials with which they are made and the defects of construction are drawbacks on their perfect working; but in the Compound Microscope we have an instrument working up to the theory of its construction. It does actually all that could be expected from it, upon a correct theory of the principles upon which it is constructed. Nevertheless, this instrument did not come perfect from its inventor's hands. Its principles were understood by the earlier microscopic observers in the seventeenth and eighteenth centuries, but there were certain drawbacks to its use, which were not overcome till the commencement of the second quarter of the present century.

These drawbacks depended on the nature of the lenses used in its construction. The technical term for the defects alluded to are *chromatic* and *spherical aberration*. Most persons are acquainted with the fact that, when light passes through irregular pieces of cut glass—as the drops of a chandelier,—a variety of colours is produced. These colours, when formed by a prism, produce a coloured image called the *spectrum*. Now, all pieces of glass with irregular surfaces produce, more or less, the colours of the spectrum when light passes through them; and this is the case with the lenses which are used as object-glasses for Microscopes. In glasses of defective construction, every object looked at through them is coloured by the agency of this property. The greater the number of lenses used in a Microscope, the greater, of course, is the liability to this colouring. This is chromatic aberration; and the liability to it in the earlier-made Compound Microscopes was so great that it destroyed the value of the instrument for purposes of observation.

Again, the rays of light, when passing through convex lenses, do not fall—when they form a picture—all on the same plane; and therefore, instead of forming the object as presented, produce a picture of it that is bent and more or less distorted. This is spherical aberration, and a fault which was liable to be increased by the number of glasses, in the same way as chromatic aberration. This defect also is increased in Compound Microscopes; and formerly, the two things operated so greatly to the prejudice of this instrument that it was seldom or never used.

Gradually, however, means of improvement were discovered. These defects were rectified in telescopes; and at last a solution of all the difficulties that beset the path of the Microscope-maker was afforded by the discoveries of Mr. Joseph Jackson Lister, a gentleman engaged in business in London, who, in a paper published in the *Philosophical Transactions* for 1829, pointed out the way in which the Compound Microscope could be constructed free from chromatic and spherical aberration. This is done by such an arrangement of the lenses in the object-glass, that one lens corrects the defects of the other. Thus, in object-glasses of the highest power, as many as eight distinct lenses are combined. We have, first, a triplet, composed of two plano-convex lenses of crown-glass, with a plano-concave of flint-glass between them. Above this is placed a doublet, consisting of a double convex lens of crown, and a double concave one of flint-glass. At the back of this is a triplet, which consists of two double convex lenses of crown-glass, and a double concave one of flint placed between them. Such are the combinations necessary to correct the defects of lenses when employed in Compound Microscopes.

It is this instrument, then, which is most commonly employed at the present day, and to which we are indebted for most of the recent progress in microscopic observation.

In using the Microscope, a great variety of accessory apparatus may be employed to facilitate the various objects which the observer has in view. As this is a book for beginners, we shall only mention a few of these.

Microscopes are generally supplied with small slips of glass, three inches long and one inch wide. These are intended to place the objects on which are to be examined. They are either used temporarily or permanently with this object in view,

Fig. 8. Forceps.

and are called *slides*. When used temporarily, an object, such as a small insect, or part of an insect,

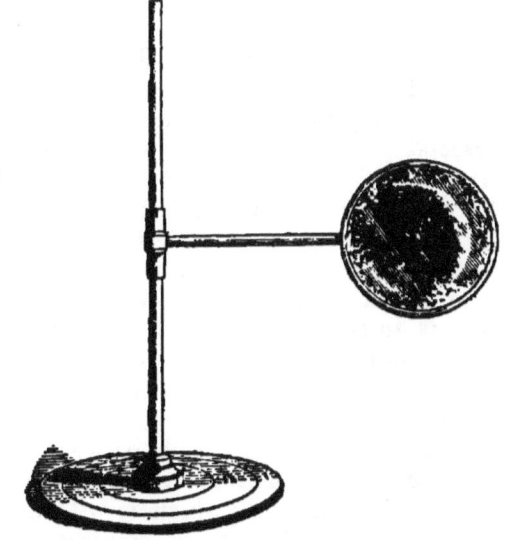

Fig. 9. Bull's-eye Condenser.

is placed upon the middle of it; and it may be either placed immediately upon the stage at the proper distance from the object-glass, or a drop of water may be placed on the slide, and a piece of thinner glass placed over the object. This is the most convenient arrangement, as you may then tilt your Microscope without the slide or object falling off.

Objects, when placed under the Microscope, are of two kinds—either *transparent* or *opaque*. When they are opaque, they may either be placed upon the slips of glass, or put between a small pair of *forceps* (fig. 8), which are fixed to the stage of the Microscope, and the light of a window or lamp allowed to fall upon them. This is not, however, sufficient, generally, to examine things with great accuracy; and an instrument called a *condenser* (fig. 9) is provided for this purpose. It consists merely of a large lens, which is sometimes fixed to the stage, or has a separate stand. Its object is to allow a concentrated ray of light to be thrown on the opaque object whilst under the object-glass of the Microscope. This is called viewing objects by *reflected light*.

Transparent objects, on the other hand, are viewed by *transmitted light*, reflected from the plane or concave surface of the mirror beneath the stage. The object of this mirror, which is called the *reflector*, is to catch the rays of light and concentrate them on the object under the Microscope. The rays of light thus pass through the object, and its parts are seen much more clearly.

Another convenient piece of apparatus is an *animalcule cage*. This consists of a little brass box, inverted, to the bottom of which is attached a piece of glass. Over this, again, is placed a lid or cover, with a glass top. The cover can be made to press on the glass beneath, and an object being placed between the two glasses, can be submitted to any amount of pressure thought necessary. (Fig. 10.) This is a very important instrument for examining minute crustacea, animalcules, zoophytes, and other living and moving objects, especially when they live in water.

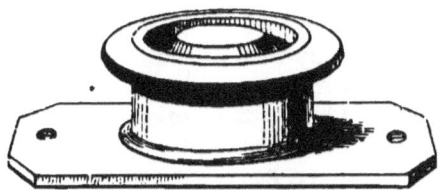

Fig. 10. Animalcule Cage or Live Box.

In the use of the cage and the slide, care must be taken not to break them by turning the object-glass down upon them. It is sometimes a difficult thing, when the object-glass has a focus of not more than a quarter or eighth of an inch, to adjust it to exactly the point at which the object is best seen, by means of the coarse handles on the rack-work. For this reason the Microscope has been provided with a *fine adjustment,* by which the object-glass is moved down on the object in a much slower and more gradual manner, and the destruction of an expensive objective glass is often thus prevented.

The picture of the object brought to the eye in the Compound Microscope is always the wrong end upwards. That is, the picture is always the reverse in the Microscope to what it is with the naked eye. You need constantly to be aware of this, especially if you are going to dissect an object under the Microscope, as your right hand becomes left, and your left right. The observer, however, soon gets accustomed to this, and it creates no difficulty ultimately. But science constantly attends on the

Microscope, and ministers to its slightest defects. A little instrument called an *erector*, composed of a lens which reverses the picture once more, is supplied by the optician, and can be had by those who practise the refinements of microscopic observation.

It is a good plan to make drawings of all objects examined, or at any rate those which are new to the observer. A note-book should be kept for this purpose, and what cannot at once be identified by the object, may afterwards be so by the drawing. All persons, however, have not the gift of drawing, and for those who need assistance in this way, the *camera lucida* has been invented. This instrument is applied to the tube of the Microscope when placed at right angles with the stem, in such a way that a person looking into it sees the object directly under his eye, so that he may easily draw its form on a piece of paper placed underneath. (Fig. 11.) Some little practice is, however, necessary before the observer can obtain satisfactory results with this instrument. It is absolutely essential that the eye should be so placed that, while one part of the pupil receives the rays from the reflecting surface

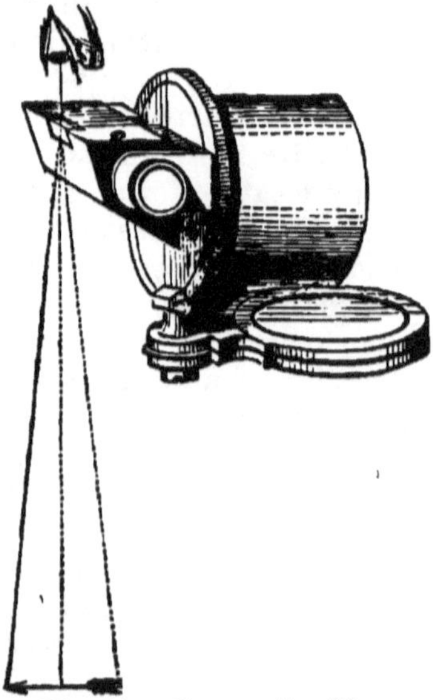

Fig. 11. Camera Lucida.

of the prism, the other sees the paper below with the image clearly depicted upon it. Dr. Beale strongly recommends the neutral lens glass reflector in preference to the Wollaston *camera lucida*. It is also much less costly. (Fig. 11A.) This consists of a short tube falling upon the eye-piece, with a piece of neutral lens glass placed at such an angle that, whilst the image of the object is reflected upwards, the paper below can be distinctly seen. (The price of this form of *camera lucida* is about four or five shillings.) Success in the use of the camera depends very much on the arrangement of the light.

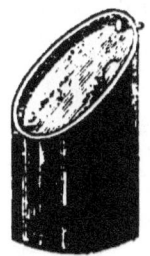

Fig. 11A.

If the image is too strongly illuminated, the paper will hardly be visible; and, on the contrary, if the paper and pencil are too bright, the image is indistinct. A little practice will enable the observer to overcome both difficulties: this he will have attained when he can see the image and paper with equal distinctness.

Another instrument which will be found of considerable service even to the beginner with the Microscope, is a *micrometer*. This is an instrument for measuring the size of objects observed. Exaggerated notions about the smallness of objects are very prevalent; and as it is almost impossible to say accurately how small an object is without some means of measuring, a Micrometer becomes essential where accuracy is desired. This is effected by having some object of known size to compare with the object observed. The most convenient instrument of this kind is a glass slide, on which lines are drawn the hundredth and thousandth of an inch apart. If this slide, or *stage micrometer* as it is called, is placed on the stage, the divisions may be traced on the paper in the same way as the outline of an object: the dimensions of the latter can now be ascertained. Care must, however, be taken that the magnifying power is the same in both cases.

Amongst the accessory apparatus are various arrangements for concentrating the light on the objects which are placed for examination under the Microscope. One of these combinations is called the *achromatic condenser*. This consists of a series of lenses, which are placed between the mirror and the stage, and which may consist of an ordinary object-glass. The stages of the larger kinds of Microscopes are fitted up with a screw or slide, by which the condenser can be fastened beneath and adjusted to the proper focus for throwing light on the object examined.

The illumination of opaque objects by means of the bull's-eye condenser is sufficient when only the lowest powers are used; but when any objective of less than inch-and-half focus is used this method of illumination is not satisfactory, and a form of reflector called a *Lieberkühn* will be found to be a welcome addition to the Microscope. This instrument consists of a concave silvered speculum with a central aperture of the diameter of the front lens of the objective: a short tube is attached to the convex surface of the reflector, which slides over the object-glass. The action of the *Lieberkühn* will be easily understood from the following diagram: a represents the objective with the *Lieberkühn in situ;* b, the concave reflector; c, a stop for the purpose of preventing any direct light entering the objective (a small disk of black paper attached to the slide is generally sufficient); $d, d,$ rays of light from the mirror; $e, e,$ reflected rays converging to a focus at f (the object). To obtain the full effect of this mode of illumination the mirror should be placed a little out of the axis of the tube of the Microscope. By this method an oblique beam of light is thrown on the *Lieberkühn,* and the light from it is reflected unequally upon the object; thus producing the light and shade so necessary for the proper definition of an object. (The cost of a *Lieberkühn* varies from 6s. to 15s.; those for low powers costing more than those for the higher.)

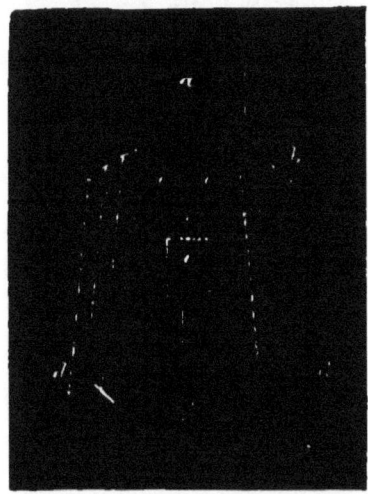

Fig. 11B. Diagram illustrating the action of a "Lieberkühn." *a,* object-glass; *b,* a concave silver reflector; *c,* a black spot ("dark well"); *d, d,* rays of light; *e, e,* the same reflected and brought to a focus at *f.*

The details of many transparent objects are much more distinctly seen when examined by light transmitted by the object only. This is called black ground illumination, and can be obtained in several ways. With a very low power the light can be reflected with sufficient obliquity if the mirror is thrown out of the axis; but much better effects are obtained when a hemispherical lens with a central black stop, called a "spot lens," is placed beneath the object. The accompanying figure will explain its action:— *a* is "spot lens"; *b,* a brass tube in which it is mounted (this is fitted into a larger tube fitted to the short tube attached to the lower surface of the stage: by sliding this up or down, the proper distance from the object is obtained); *c,* parallel lines from mirror; *d*, the same rays made to rapidly converge by passing through the lens, and come to a focus at *e*; and if the focal length of the objective is greater than the distance between the object and the point *e*, the object will be illuminated, and the field appear perfectly dark.

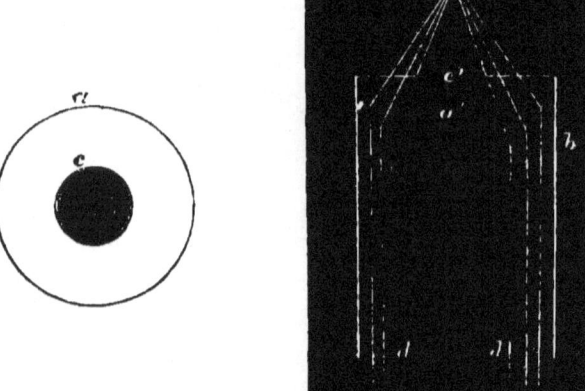

Fig. 11c.

a, "Spot Lens," front view; *c*, blackened concavity of ditto; *a'*, section of "Spot Lens" in its fitting, *b*; *c'*, central stop; *d d*, parallel rays of light converging to a focus at *e*.

Having said thus much with regard to apparatus, we will now give some directions for the use of the Microscope under ordinary circumstances. The Microscope may be either used by the light of the sun in the daytime, or at night by some form of artificial light. It is best used by daylight, as artificial light is likely to tire the eyes.

Having determined to work by daylight, some spot should be selected near a window, out of the direct light of the sun, in which to place a small, firm, steady table. On this the Microscope should be placed, and the object-glass should be screwed on to the tube. The mirror should be then adjusted so as to throw a bright ray of light on to the object-glass. The eye-piece having been previously placed at the top of the tube, the Microscope is now ready to receive a transparent object. If the object to be examined is an animalcule, it may be conveyed to the animalcule-cage by means of a glass tube, called a *pipette* or *dipping-tube* (fig. 12), which should be dipped into the water where the object is contained, with the finger covered over the upper orifice, so that no air can escape. By taking the finger off when the tube is in the water, the fluid will rush into the tube, and with it the object to be examined. The finger is again applied to the top of the tube, and the fluid obtained conveyed to the animalcule-cage. Only such a quantity of the water should be allowed to fall out of the tube on to the cage as will enable the observer to put on the cover of the cage without pressing

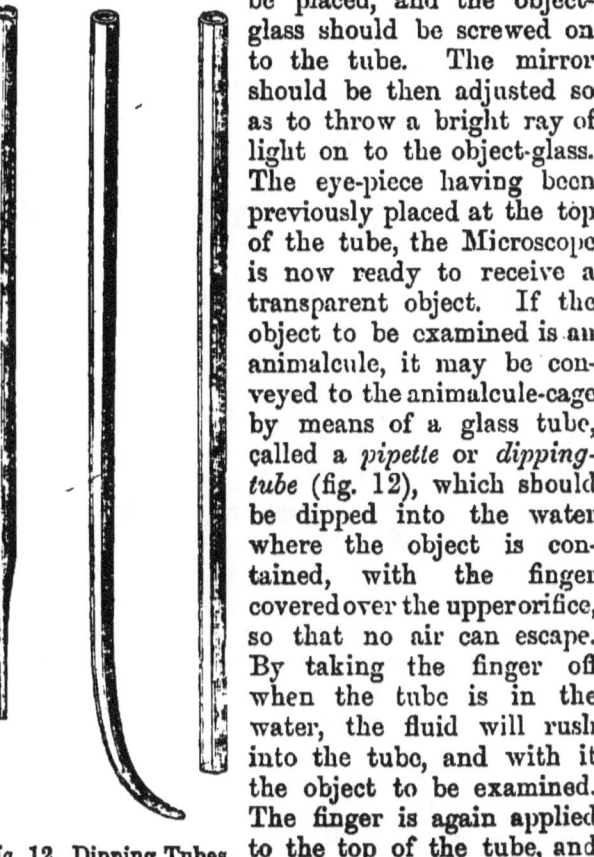

Fig. 12. Dipping Tubes.

the fluid out at the sides of the cage. If the water is thus allowed to overflow, it runs over the glasses of the cage, and thus obscures vision. An object or objects having been thus placed in the cage, it is conveyed to the stage, and placed in such a position that the ray of light passing from the mirror to the object-glass may pass through it. This having been done, the observer must now place his eye over the eye-piece, and use the screw in the tube, and move the object-glass downwards until he gets a clear view of objects moving in the water. This is called *focussing*. The glass may then be moved up or down, in order that the best view of the object may be obtained. When the object-glass is one of high power, the *fine adjustment* may be used for this purpose. When the proper focus is obtained, the object may be moved up or down, right or left, with the hand, or by the aid of the screws which are employed in the various forms of what are called *mechanical stages*.

When objects not requiring the live-box or animalcule-cage are to observed, they may be transferred to the glass slide by aid of a thin slip of wood, or a porcupine-quill moistened at the end, or by a pair of small *forceps*. (Fig. 8.) Some transparent objects may be seen without any medium, but generally it is best to place them on the slide with a drop or two of clean water, which may be placed on it with a *dipping-tube*. When water is used, it will generally be found best to cover the object with a small piece of thin glass. Small square pieces of thin glass are sold at all the opticians' shops for this purpose. The object is then placed under the object-glass as before.

In order to render objects transparent, so that

they may be viewed by transmitted light, very thin sections of them should be made. This may be effected by means of a very sharp scalpel, or a razor. When objects are too small to be held in the hand to be cut, they may be placed between two pieces of cork, and a section of them made at the same time that the cork is cut through.

Sometimes it is found desirable to unravel an object under the Microscope. If this is the case, only a low power should be used, and the object may be placed on a glass slide, without any glass over, and two needles with small wooden handles employed,— ordinary sewing needles, with their eyes stuck in the handle of a hair pencil, will answer very well. (Fig. 14.) Even when dissection is not to be carried on under the Microscope, a pair of needles of this sort, for tearing minute structures in pieces, will be found very useful.

Fig. 14.
Dissecting Needles.

When opaque objects are to be examined, the light from the mirror may be shut off, and the aid of the bull's-eye condenser called in. The object being secured in the forceps attached to the stage (fig. 15), or laid upon a slide, the light is allowed to fall on it through the condenser. (Fig. 9.) The object-glass must be focussed in the same manner as for transparent objects, till the best distance is secured for examining it. The petals of plants, the wings and other parts of insects, with many other objects, can only be examined in this way.

Fig. 15. Stage Forceps.

Even the beginner will find it useful to keep by him some little bottles, containing certain chemical re-agents. Thus, a *solution of iodine* is useful to apply to the tissues of plants, for the purpose of ascertaining the presence of starch. This solution may be made by adding five grains of iodine and five grains of iodide of potassium to an ounce of distilled water. It turns starch blue and cellulose brown. Cellulose is the substance that forms the walls of the cells in plants. Dilute *sulphuric acid* (1 to 3) is also useful as a re-agent; if applied to cellulose previously stained with iodine, it imparts a blue or violet tint. *Strong nitric acid* turns albuminous matter a deep yellow; and when diluted (1 to 4) with water is used for separating the elementary tissues of vegetable substances either by boiling or maceration.

The *strong solution of potash* (liquor potassæ) can also be employed with advantage in softening and making clear opaque animal and vegetable substances. While using these powerful agents, great care should be taken to prevent the transparency of the object-glass becoming impaired by contact with them or by long exposure to their vapours.

CHAPTER II.

A HALF-HOUR WITH THE MICROSCOPE IN THE GARDEN.

AMONGST the objects which can be examined by the Microscope, none are more easily obtained than plants. All who have a Microscope may not be fortunate enough to have a garden; but plants are easily obtained, and even the Londoner has access to an unbounded store in Covent Garden. We will, then, commence our microscopic studies with plants. On no department of nature has the Microscope thrown more light than on the structure of plants; and we will endeavour to study these in such a manner as to show the importance of the discoveries that have been made by the aid of this instrument.

If we take, now, a portion of a plant, the thin section of an apple, or a portion of the coloured parts of a flower, or a section of a leaf, and place it, with a little water, on a glass slide under the Microscope, we shall see that these parts are composed of little roundish hollow bodies, sometimes pressed closely together, and sometimes loose, assuming very various shapes. These hollow bodies are called "cells," and we shall find that all parts of plants are built up of cells. Sometimes, however, they have so far lost their cellular shape that we cannot recognize it at all. Nevertheless, all the parts we see are formed out of cells. Cells tolerably round, and not pressed on each other, may be seen in most pulpy fruits. In fact, with a little care in making a thin section, and placing

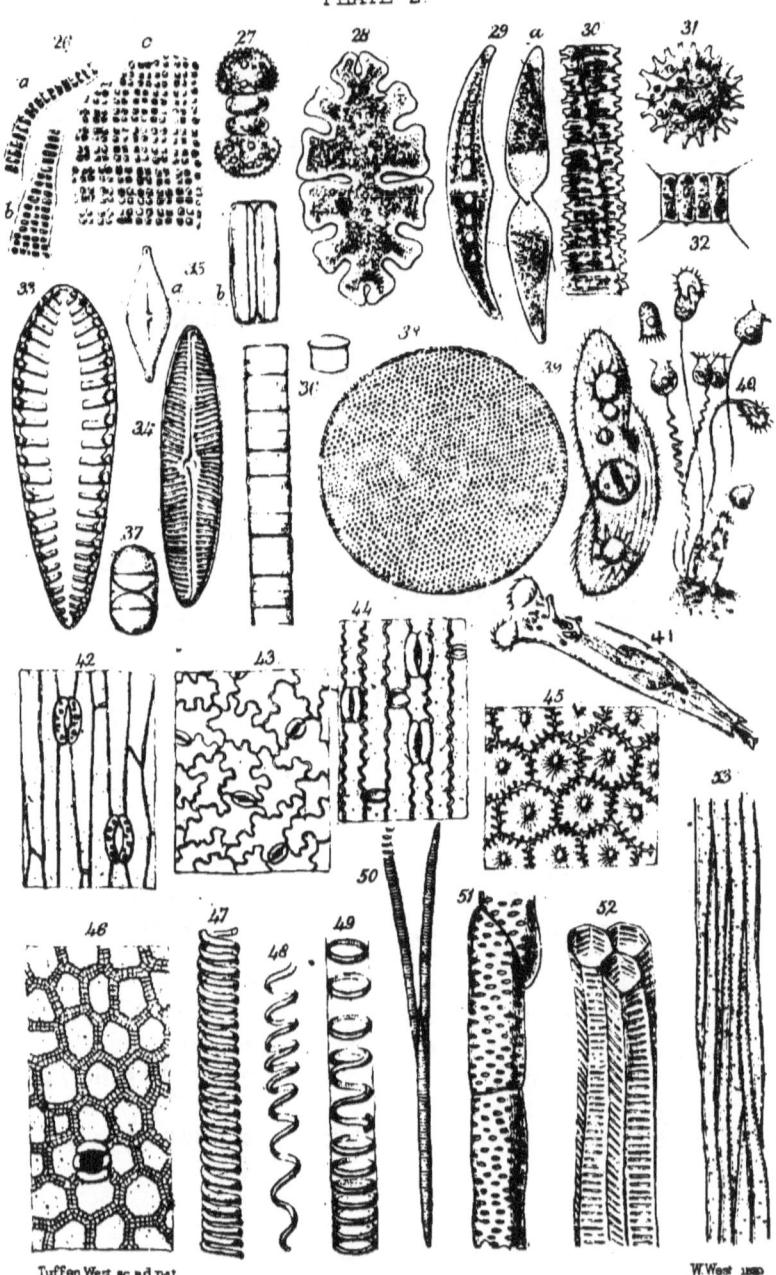

PLATE 2.

Tuffen West sc ad nat. W. West imp

London: Robert Hardwicke, 1860.

it under the Microscope, the cellular structure of plants may be observed in all their soft parts.

If, now, we take a thin section from an apple, or other soft fruit, or from a growing bud, or tuberous root, as the turnip, we shall find that many of the cells contain in their interior a "nucleus," or central spot, a representation of which is seen from the cells of an apple in figure 1 of the first plate. This nucleus is a point of great importance in the history of the cell, for it has been found that the cell originates with it, and that all cells are either formed from a nucleus of this kind, or by the division of a thin membrane in the interior of the cell, which represents the nucleus, and is called a "primordial utricle."

When the cells of plants have thus originated, they either remain free or only slightly adherent to each other, or they press upon each other, assuming a variety of shapes; they then form what is called a "tissue." When cells are equally pressed on all sides, they form twelve-sided figures, which, when cut through, present hexagonal spaces. This may be seen in the pith of most plants, more especially the common elder, which is seen at figure 2 of plate 1. Transverse slices of the stems of any kind of plant from the garden may be made by a razor, or sharp penknife, and will afford interesting objects for the Microscope.

Cells, during their growth, assume a variety of shapes, and the tissues which they form are named accordingly. Two examples of such cells will be seen in figures 243 and 244 in plate 8, where the first represent cells from the hard shell of a plum stone, and the second the thin cells from the outside of the seed of the guelder rose. Sometimes the cells are very much elongated, or they unite together to form an elongated tube; the tissue thus formed is called "vascular tissue;" but where the cells retain their primitive form, it is called "cellular tissue." A very interesting form of the latter is the "stellate" tissue found in most water plants, and especially regularly developed in the common rush, a representation of which is given in figure 3 plate 1. The object of this tissue is, evidently, to allow of the existence of a large quantity of air in the spaces between the cells; by which means the stem of the plant is lightened, and it is better adapted for growth in water.

If the leaf of any plant is examined, it will be found that on the external surface there is a thin layer, called, after the thin external membrane in animals, the "epidermis." This layer is composed of very minute cells—smaller than those in other parts of the plant, and when placed under the Microscope, presents a variety of forms of cellular tissue. The form of epidermal cells from various plants is seen in figure 42 and the following figures in plate 2. There is found in this layer a peculiar organ which exists on the outside of all parts of plants, and which demands attention. In the midst of the tissue, at very varying distances, are placed little openings, having a semilunar cell on each side. These openings are called "stomates," and can be well seen in the leaf of the hyacinth, which is shown in figure 42, where the cells of the epidermis are transparent; but the little cells which form the stomate are filled with green colouring-matter. The stomates vary very much in size and in numbers. They are found in larger numbers on the lower than on the upper side of leaves. In the common water-cress they are very small, as seen in figure 43, plate 2, and the cells of the epidermis are sinuous. The stomates are found on all plants having an epidermis.

In figures 44 and 46 they are represented from the wheat and the aloe. In the latter plant the cells of the cuticle are very much thickened. They can also be seen on the cuticle of the fruit, as shown from the holly in figure 241, plate 8, and also on the organs and petals. These form a beautiful object under the Microscope. The petal of the common scarlet geranium (*Pelargonium*) affords a beautiful instance of the way in which the cells of plants become marked, by their peculiar method of growth. This is illustrated in the cells of the common red-flowered geranium at figure 45, in plate 2.

The vascular tissue of plants is either plain or marked in its interior. If we examine the ribs of leaves, the green stems of plants, or a longitudinal section of wood, elongated fibres, lying side by side, are observed, as is seen in the case of the elder, at figure 53, plate 2. This is what is called "ligneous" or "woody" tissue, and the greater part of the wood and solid parts of plants are composed of this tissue. Such tissue is seen upon the shoots of the young vine in figure 249, plate 8. The fibres mostly lie in bundles, and are divided from each other by cellular tissue. This latter, in the woody stems of trees, constitutes the "medullary rays," which are seen in transverse sections of stems, extending from the pith to the bark. The difference observable in the distribution of the woody fibres and the medullary rays renders the examination of transverse sections of the stems of plants a subject of much interest; figure 54 and the following figures in plate 3, present the appearances of thin sections of various kinds of wood (figures 54, 55, 56, 57, plate 3). In the transverse sections of stems of most plants, large open tubes are observed. This is seen in the case of the oak, figured at figure 55, plate 3. These are called "ducts." Such ducts may be well observed in the transverse section of the common radish, as seen at figure 51, plate 2, and in other roots. These ducts are often marked by pores, or dots, and are hence called "dotted ducts." These dots are the result of deposits in the interior of the tube of which the duct is formed, and a great variety of such markings are found in the interior of vascular tissue. One of the most common forms of marked vascular tissue is that which is called glandular woody tissue, of which a figure is given at 54, plate 3. This kind of tissue is found in all plants belonging to the cone-bearing, or fir tribe of plants. In order to discover it, recourse need not be had to the garden for growing plants, as every piece of furniture made of deal wood will afford a ready means of obtaining a specimen. All that is necessary to observe the little round disks with a black dot in the middle is to make a thin longitudinal section of a piece of deal, and place it under a half or quarter-inch object-glass, when they will be readily apparent. The application of a drop of water on the slide, or immersing them in Canada balsam, will bring out their structure better.

If we take the leaf-stalk of a strawberry, or of garden rhubarb, and make a transverse section all round, nearly to the centre of the stalk, the lower part will at last break off, but be still held to the upper by very delicate threads. If we examine these threads, we shall find that they are fibres which have been left by the breaking of the vessel in which they were contained: such fibres are seen at figure 48, plate 2. These vessels are called "spiral vessels," and are found in the stems and leaves of many plants. They are seen rolled up as found in the garden rhubarb, at figure 47, plate 2. Sometimes these vessels are found branched, as in the common chickweed, which is seen at figure 50, plate 2. This arises from two spires coming in contact with each other, and adhering. Occasionally the spiral fibre breaks, or is absorbed at certain points, leaving only a circular portion in the form of a ring, as seen in a vessel from the root of wheat at figure 49, plate 2. Such vessels are called "annular," and may be observed in other roots besides those of growing wheat, as in the leaves of the garden rhubarb. A modification of this kind of tissue is seen in the stems and roots of ferns, in which the vessel assumes a many-sided form. This kind of tissue is called "scalariform," or ladder-like, and is seen in figure 52, plate 2. Sometimes the spiral fibre is free. This is represented at figure 250, plate 8, from the testa of the seed of the wild sage.

The bark as well as the wood of trees affords the same appearance under the Microscope. If a piece of the bark of any plant be examined by means of a very thin transparent section, and placed upon a slide, and put under an inch or a half-inch object-glass, the structure of the bark may be easily seen. On the outside of all is the cuticle, or epidermis, and under this lie two layers, composed, like the cuticle, of cellular tissue; but the inner layer, before we come to the wood of the stem, is composed of woody tissue. The cellular layer, next the woody one, is often developed to a very great extent, and then constitutes what we know by the name of cork. The bark from which corks are made is obtained from an oak tree which grows in the Levant. If we make a very thin section of a cork, its cellular structure can be easily made out. The cells are almost cubical, and when submitted to the action of a little solution of caustic potash, they may frequently be seen to be slightly pitted. This is represented from cork in figure 59, plate 3.

Many of the structures which are described above may be seen in common coal; thus proving most satisfactorily that this substance has been formed from a decayed vegetation. A transverse and a longitudinal section of coal is shown at figures 60 and 61, plate 3. The examination of coal, however, is by no means an easy task, and the hands and fingers may be made very black, and the Microscope very dirty, without any evident structure being made out. Some kinds of coal are much better adapted for this purpose than others. Sections may be made by grinding, or coal may be submitted to the action of nitric acid till it is sufficiently soft to be cut. The amateur will not find it easy work to make sections of coal; but should he wish to try, he may fasten a piece on to a slip of glass with Canada balsam, and when it has become firmly fixed, he may rub it down on a fine stone till it is sufficiently thin to allow its structure to be seen under the Microscope. Coal presents both vascular and cellular tissue. The vascular tissue is, for the most part, of the glandular woody kind; thus leading to the inference that the greater portion of the vegetation that supplied the coal-beds belonged to the family of the firs.

The external forms of the tissues of plants having been examined, we are now prepared to regard their contents. In the interior of the cells forming the roots and the growing parts of plants will be observed a number of minute grains, generally of a roundish form. If we make a thin slice of a potato, these granules may be very obviously seen, lying in the interior of the cells of which the potato is composed, as seen at figure 64, plate 3. If we now take a drop of the solution of iodine, and apply it to these cells full of granular contents, we shall find that the granules assume a deep-blue colour. This is the proof that they are starch; and as far as we at present know, no other substance but starch has the power of assuming this beautiful blue colour under the influence of iodine. We have thus a ready means at all times of distinguishing starch. The grains of starch are of various sizes and shapes. The starch of the flour of wheat has a round form, and varies in size; that of the oat is characterized by the small granules of starch adhering together in globular shapes. When these globules are broken up, the grains appear very irregular. Grains of wheat starch and oat starch are seen in figures 62 and 63, plate 3. In the arrow-root called "Tous les Mois," the grains of starch are the largest known, and, like those of the potato, they look as if composed of a series of plates laid one upon the other, gradually becoming smaller to the top. This is seen at figure 65, plate 3. These lines do not, however, indicate a series of plates, but appear more like a series of contractions of a hollow vesicle or bag. This vesicular appearance of starch may be made apparent by gently heating it, after moistening, over a spirit-lamp on a glass slide, or by dropping on it a drop of strong sulphuric acid. This action of the starch-granule appears to be due to the fact that the starch is converted into gum by the action of the heat on the sulphuric acid. Sago and tapioca are almost entirely composed of starch, and may be easily examined under the Microscope. Granules of sago are represented in figure 67, and those of tapioca at figure 68; they are readily distinguished by their size. The starch granules are insoluble in water, but they are easily diffused through it; so that by washing any vegetable tissue containing starch, with water, and pouring it off and allowing it to stand, the starch falls to the bottom. This may be done by bruising the vegetable tissue in a mortar, and then throwing it into cold water. The tissue falls to the bottom, and the starch is thus suspended in the water. In this way the various kinds of starches may be procured for microscopical examination. The granules of starch have frequently a little black irregular spot in their centre. In the starch of Indian corn it assumes the form of a cross, which is seen at figure 66, plate 3. Starch is a good object for the use of the polarizing apparatus, which can be applied to most compound Microscopes. The grains of starch, under the influence of polarized light, become coloured in a beautiful and peculiar manner, permitting of great variation, as in the case of all polarized objects.

If we take a little of the white juice from the common dandelion, and put it under the Microscope, we shall often see, besides the globules of caoutchouc which make the juice milky, crystals of various forms. Such crystals are called by the botanist "raphides,"—signifying their needle-like form. They arise from the formation and accumulation of insoluble salts in the fluids of the plant. They are seen in various plants, and under very different circumstances. Beautiful needle-like crystals can be seen in the juice of the common hyacinth, represented at figure 69; the juice may be obtained by pressing. A question has been raised as to whether they are always formed *in* the cell. They are mostly found lying in the cell, as in the leaves of the common aloe, seen at figure 70, plate 3: they may also be found in the tissues of the common squill, and in the root of the iris. If a thin

PLATE 3

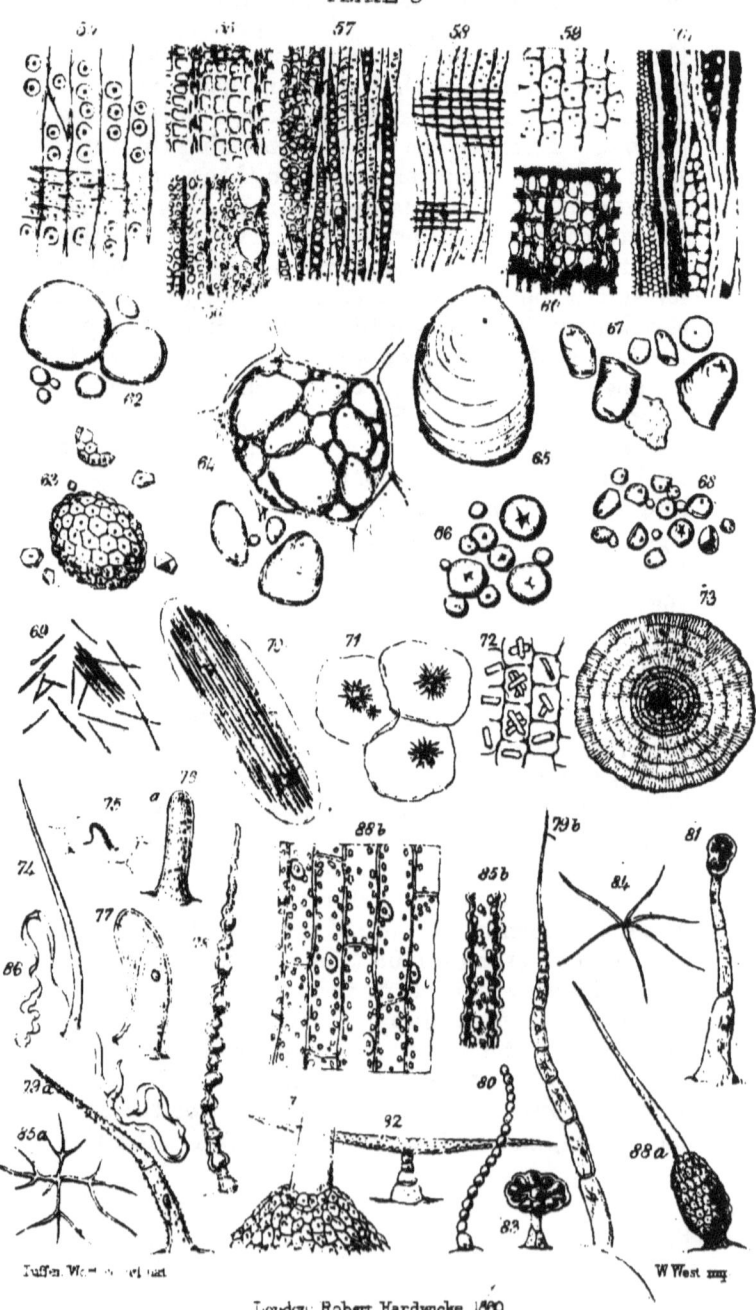

London: Robert Hardwicke, 1860

section of the brown outer coat of the common onion is made, small prismatic crystals are observed. These are represented at figure 72, plate 3. Sometimes several of these crystals unite together around a central mass, forming a stellate body. These bodies have been called "crystal glands," but they have no glandular properties. They may be seen in the root and leaf-stalk of common rhubarb, and may be easily observed in a bit of rhubarb from a spring tart. From such a source, the drawing was made at figure 71. These crystals are mostly formed of oxalate of lime. They are constantly found in plants producing oxalic acid. The gritty nature of rhubarb root arises from the presence of oxalate of lime. Sometimes the oxalate of lime assumes a round dish-like form. Such forms are seen in plants belonging to the cactus family. A circular crystalline mass, as seen in a common cactus, is represented at figure 73.

Other substances, besides oxalate of lime, are found crystallized in the interior and on the surface of plants. Crystals of sulphate of lime have been found in the interior of cycadaceous plants. Carbonate of lime is found in crystals on the surface of some species of *Chara,* or stonewort. There is a shrub not uncommon in gardens, known by the name of *Deutzia scabra,* on the under surface of the leaves of which there are beautiful stellate crystals of silica. The best way of seeing these is to put the leaf under the Microscope, and to examine it by the aid of reflected light.

Sugar and honey assume a crystalline form, and may be known by the shape of their crystals. At figure 238, plate 8, a crystal of honey is represented; it is thinner and smaller than the crystal of cane sugar represented at figure 239. Honey is sometimes adulterated with sugar. Under these circumstances the sugar crystal loses its definite outline, and assumes the form seen at figure 240.

The external surface of the parts of all plants will afford a rich field of amusement and instruction to the microscopic observer. The cuticle, or epidermis, of which we have before spoken, has a very varied structure, and contains the little openings (stomates) before described. The cuticle, which, in a large number of cases, is smooth, becomes elevated in some instances, and forms a series of projections, which, according to their form, are called " papillæ," " warts," " hairs," " glands," and " prickles." The papillæ are slight elevations, consisting of one, two, or more cells; the warts are larger and harder; whilst the hairs are long, the glands contain a secretion, and the prickles are hard and sharp. For examining the form and growth of these hairs, the flowers of the common pansy (heart's-ease) afford a good object. Some of the projections are merely papillæ, as in the case of the kind of rudimentary hair represented in figure 75, plate 3; others are found longer, and more like hairs, as seen in figure 76; whilst others are long, and, the sides of the hair having contracted, they assume the appearance of a knotted stick, as seen in the hair from the throat of the flower of the pansy, at figure 78. The family of grasses, wheat, barley, oats, and other forms, are favourable subjects for the examination of simple hairs, or hairs composed of a single elongated cell. At figure 74, a single hair is given from a common grass. All that is necessary to be done, in order to see these hairs, is to take any part of the plant where they are present, and to slice off a small portion with a sharp penknife or razor, and place it under the Microscope. They may be either examined dry, or a little water may be added, and a piece of thin glass placed over them on the slide. Hairs are frequently formed of several cells. On the white dead-nettle the hairs are composed of two cells, as seen in figure 79a. The nucleus, or cytoblast, is often seen in these, and is represented in figures 76, 77, and 79, plate 3. On the common groundsel hairs may be seen, composed of several cells, each cell containing a nucleus, as at figure 79b. Hairs like a string of beads are found on the pimpernel and sow-thistle, which last will be found in figure 80, plate 3. Occasionally hairs become branched. Thus, on the leaf of the common chrysanthemum the hairs present the form of the letter T. This hair is represented at figure 82. On the under-surface of the leaves of the common hollyhock hairs are seen with several branches, giving them a stellated appearance, as seen at figure 84. The common lavender is covered with stellate hairs, as seen at figure 85a. These hairs may be examined as opaque or transparent objects, when immersed in a little glycerine. The hair of the tobacco plant presents a peculiar knobbed appearance. The presence of these hairs is a test of the purity of tobacco. It is shown in figure 81. The verbena has rosette-shaped hairs, as in figure 83. Sometimes hairs are covered over with little dots, which are supposed to be deposited after the growth of the cells of the hair. Such hairs may be seen in the common verbena, and are represented at figure 85b. Hairs are sometimes loose and long, as in the white poplar, seen at figure 86. Occasionally an elevation, consisting of several cells, is formed at the base of a hair. These are shown in figure 87. When these cells contain a poisonous secretion, which is transmitted along the tube of the hair, the hair is called a glandular hair, or sting. Such are the hairs of the common stinging-nettle, represented at figure 88a.

The hairs constituting the down or "pappus" of compositous plants assume a variety of forms. The seed or fruit of the common groundsel has a beautiful crown, given at figure 245, in plate 8. The pappus of the dandelion appears notched, as seen at figure 246. The burdock has a cottony hair, while the goatsbeard is like a feather,—both of which are represented respectively in figures 247 and 248.

If a hair is examined in its growing state, with an object-glass of one quarter of an inch focus, a movement of the particles in its interior is often observed. This is easily seen in the hairs around the stamens of the common Spiderwort (*Tradescantia Virginica*). Such movements are very common in the cells of water plants. One of those most commonly cultivated in aquavivaria at the present day, the *Valisneria spiralis*, affords the best example of this interesting phenomenon. In order to observe this movement, a growing leaf of the valisneria should be taken, and a longitudinal slice should be removed from its surface, by means of a sharp penknife or razor. The slice, or the sliced part left on the leaf, should now be put on a slide, a drop or two of water added, and covered with a thin piece of glass, when, after a little time, especially in a warm room, the movement will be observed. This movement takes place in the little particles around the sides of the cells represented in figure 88b, plate 3. It may also be seen in the leaves of the new water-weed (*Anacharis alsinastrum*), the frogbit, the rootlets of wheat, in the family of charas, and in the cells of many other water plants. In examining some species of *Chara*, the external bark, or rind, should be removed from the cells, or the movements will not be seen. This movement seems dependent on the internal protoplasmic matter, or "primordial utricle," which is contained in many cells, and which, in these cases, is spread over the interior of the cell. It is, however, capable of contraction, and when the plants are exposed to cold, the utricles contract and prevent the movement of the contents in the interior. It is, apparently, the extension of this substance beyond the walls of the cell which constitutes the little hairlike organs called "cilia," which are constantly moving, and by the aid of which the spores of some plants effect rapid movements. Such organs are found in the *Pandorina Morum* and *Volvox globator*, moveable plants represented at figures 13 and 14, plate 1. The effect of these cilia in producing the movements of plants is well seen in the *Volvox globator*, which, on account of its rapid movements, was at one time regarded as an *animalcule*, but it is now regarded as a plant. Cilia are, however, more frequently met with in the animal kingdom. They are seen in the drawing of *Plumatella repens*, at *a*, in figure 163 of plate 6.

Amongst the parts of plants which can alone be investigated by the Microscope are the stamens. These organs are situated in the flower, between the petals and the pistil, and usually consist of a filament, or stalk, with a knob or anther at its top. If the anther is examined, it will usually be found to consist of two separate valves, or cases, in each of which is contained a quantity of powder, or dust, called "pollen." The walls of these valves are worth careful examination under the Microscope, on account of the beautifully-marked cellular tissue of which their inner walls consist. The cells of this tissue contain in their interior spiral fibres similar to those which have been described as present in certain forms of vascular tissue. In the anthers of the common furze the fibres are well marked, and are represented in figure 118, plate 5; in the common hyacinth they are larger, and frequently present, in their intercellular spaces, bundles of raphides, as seen at Figure 119. In the white dead-nettle the fibre is irregularly deposited, as at figure 120. In the anthers of the narcissus, given at figure 121, the cells are almost vascular in their structure, and present the same appearance as those described under the head of annular ducts. The reader should compare figure 121, plate 5, with figure 49, plate 2. In the crown imperial the fibres of the cells radiate from a central point in a stellate manner, as at figure 122.

When the anther-cases have been examined, a little of the dust may be shaken on to a slide, and examined as an opaque or a transparent object. Each species of plant produces its own peculiar form of pollen. These little grains are actual cells. They are the cells of plants which in their position in the anther will not grow any further. They are destined to be carried into the pistil, where, meeting with other cells, they furnish a stimulus to their growth, and the embryo, or young plant, is produced. The history of the development of these cells, as well as of those in the interior of the pistil, is a very interesting one, and is one of those subjects of investigation which has been created by the aid of the Microscope. The pollen grains vary in size as well as form. They are frequently oval, as seen in figure 123, plate 5. In the hazel and many of the grasses they are triangular. Those from the hazel are represented at figure 124. In the heath they are tri-lobed, as at figure 125; in the dandelion, and many of the compositous order of plants, they are beautifully sculptured, as seen at figure 126. In the passion-flower, three rings are observed upon them, as though they had been formed with a turner's lathe—figured at 127. In the common mallow, they are covered all over with little sharp-pointed projections, like a hand-grenade. These are represented at figure 128. The microscopic observer should make himself acquainted with the forms of pollen grains, as, on account of their small size and lightness, they are blown about in all directions, and may be found on very different objects from those in which they have been produced. Some absurd mistakes have been committed by confounding pollen grains with other forms of organic matter. Thus, pollen grains in bread were regarded as bodies connected with the production of cholera.

The pistil, which is the central organ seated in the midst of the stamens in the flower of plants, will afford a great variety of interesting points for examination with the Microscope. In the earliest stages of the growth of the pistil, thin sections of it may be made, and the position of the ovules observed. In the ovule will be found the embryo sac, a central cell, which, on being brought in contact with the pollen grain, grows into the seed. The seed contains the embryo, or young plant. In most plants this is sufficiently large to be seen by the naked eye; but it may, nevertheless, be examined with advantage by a low microscopic power The seed is covered on the outside with a membrane, which is called the "testa." This membrane is often curiously marked, and the whole seed may be examined as an opaque object with the low powers of the Microscope. In order to do this, the light must be shut off from the mirror, and, the object being placed on the stage, a pencil of light should be thrown upon it by the aid of the bull's-eye condenser. If a seed of the red poppy be now examined, it will be found to have a uniform shape, and to be reticulated on its surface, as seen at figure 129, plate 5. The seed of the black mustard exhibits a surface apparently covered with a delicate network, seen at figure 130. Some seeds have deep and curved furrows on their surfaces, such as exhibited in figure 131. The great snapdragon has a seed covered with irregular projecting ridges, having a granuled appearance, represented at figure 132. The seed of the chickweed presents a series of blunt projections, as in figure 133. In the various forms of umbel-bearing plants, the seeds adhere to the fruit, and the fruit is commonly called the "seed." Such are caraway, coriander, dill, and anise seeds. The plants of this family are very common weeds in our gardens and fields, and may be easily procured for microscopic examination. Some of these fruits are covered over with little hooks, seen at figure 134, whilst others present variously-formed ridges and furrows, which are amongst the best means for distinguishing these plants the one from the other.

PLATE 4

CHAPTER III.

A HALF-HOUR WITH THE MICROSCOPE IN THE COUNTRY.

A COMPOUND Microscope is not easily conveyed and put up in the fields, but the produce of the roads and waysides may be easily brought to the Microscope at home. No one who has a Microscope should walk out into the country without supplying himself with a few small boxes, a hand-net, and three or four small bottles, in order to bring home objects for examination. The dry produce, which may be put into boxes, is of a different character from that which may be conveyed home in bottles. We shall, therefore, first direct attention to the minute forms of mosses, fungi, lichens, and ferns, which may be collected in boxes; premising, however, that many members of these families may be found without going into the country to seek for them. The cheese in the pantry, and the decayed parts of fruits, and objects covered with mould, are good subjects for microscopic examination.

Amongst the minuter plants and animals whose true nature can only be detected by the Microscope many are composed of a single cell, whilst others, like higher plants and animals, are formed by the union of a large number of cells. The greater proportion of the one-celled, or unicellular plants, as they are called, are found in water; but some are found on moist rocks, stones, and old walls. Amongst these there is one of exceedingly simple structure, called gory dew (*Palmella cruenta*). This plant appears as a red stain upon the surface of damp objects. If a little of this red matter is scraped off the object to which it is attached, and placed under the Microscope, it will be found to consist of a number of separate minute cells, as represented at figure 89, plate 4. This plant belongs to the same family as the red-snow plant, and there are a number of forms of these minute organisms, which, on account of their rapid growth and red colour, have given rise to alarming apprehensions, in former times, when their true nature was imperfectly understood. One of them attacks bread, and gives to it the appearance of having been dipped in blood. They also attack potatoes. Of the same simple structure, but not having a red colour, is the yeast-plant, or fungus, shown at figure 90, plate 4. This plant abounds in yeast, and may also be found in porter and ale. If vinegar is allowed to stand for some time, a minute plant is developed, called the vinegar-plant. In its earlier stages of growth it exhibits elongated cells, looking like broken pieces of thread, seen at figure 91. Threads more fully developed are often seen in decomposing fluids, and upon the surface of decomposing animal and vegetable substances; such is the so-called cholera-fungus, which may be obtained by exposing damp slides to the air. They are shown at figure 92. Such plant-like threads can be collected from the air in damp and unwholesome cellars and rooms, and were at one time supposed to be connected with the production of that fearful disease, the cholera. It has been rendered, however, exceedingly probable that all these appearances are but different forms of the fungus which produces common mould, and which is known by the name of *Penicillium glaucum*. This fungus is represented at figure 95. It may be found on the surface of preserves and jellies, and consists of a mass of filaments or threads serving as its base, from the surface of which individual filaments rise up, bearing a number of minute cells, which are the spores, or reproductive organs. These are seen at figure 96.

Plants such as these, and belonging to the family of fungi, are found everywhere on the leaves of plants in the summer and autumn, forming irregular spots, of a yellow, red, or black colour. If such leaves are brought home and placed under the Microscope, they present a never-failing source of interest. The red appearance on the leaves of wheat, called the rust, is due to one of these fungi, seen at figure 93, plate 4. This appears to be an early stage of the fungus, which produces what is called mildew, and is represented at figure 94. These fungi are so common on the wheat-plant that their spores mingle with the seeds when ground into flour, and can be found, when carefully sought for, in almost every piece of bread that is examined under the Microscope. Mouldy grapes, pears, apples, and other fruits, present fungi, having the same general form as that of common mould. Such a fungus is the *Botrytis* of mouldy grapes seen at figure 96. Mouldy bread also presents a fungus of this kind. This species is called *Mucor mucedo*, and is represented at figure 97. Its spores are arranged in a globular form. A fungus not unlike the last has been described as growing in the human ear, and is figured at 98. The leaves of the common bramble present a fungus in which the spores are arranged on a more dense and elongated head. This is called *Phragmidium bulbosum*, and is represented at figure 99. The *Oïdium* which attends the blight of the vine, seen at figure 100, and the *Botrytis* which accompanies the potato disease, figure 101, are other and interesting forms of these minute parasites. The common pea is subject to a blight which is accompanied by a peculiar fungus, seen at figure 102*a*, which, when examined by a low power, presents a globular mass, surrounded by minute filaments. Under a high power the central ball is resolved into a series of little cases, containing in their interior the minute spores. These are seen at figure 102*b*. Seeds, as well as fruits, are liable to the attacks of fungi during their decay. Figure 103, Plate 4, represents a fungus found in a mould upon a common Spanish nut. This fungus looks like a red powder spread over the surface of the nut. A fungus has been described as attacking the oil-casks in the London docks: its fibres resemble threads of black silk. It is represented at figure 104. The spores are found scattered about the fibres. As we have already seen, fungi are found on the human body, and accompany certain forms of disease of the skin, more especially those of the head. In these cases the fungi insert themselves into the follicle of the hair, and introduce themselves into its structure, so that it either falls off or becomes disorganized. The fungus of ringworm, called *Achorion Schönlenii*, is given at figure 105. If the seed of wheat is allowed to germinate in a damp place, the little rootlet which it sends down will be found covered over with a minute fungus. A fungus of some interest, on account of its unusual place of growth, may be found, in autumn, attached to the roots of the common duck-weed (*Lemna minor*), seen at figure 106 — plate 4. In the same figure, at *a*, is represented a fungus of a different kind, it is parasitic within the cells, and has a bead-like appearance. It may be an earlier stage of the growth of the former.

The microscopic structure of the higher forms of fungi is not without its interest. In the fungi a very elongated form of cellular tissue frequently occurs, and in the stem of the common mushroom it will be seen to be branched, as at figure 103. The looser portions of the fibres of the mushroom, which are found in the earth at the bottom of the stem, afford even a better illustration of this structure, and is given at figure 107. The gills of the mushroom, when put under the Microscope, display a number of small projections surmounted with four round cells; these are the spores arranged in fours, and which, on that account, are called *tetraspores*. They are seen at *c*, figure 107.

In the woods, in winter time, fungi abound, and their parts may be examined under the Microscope with great interest. Amongst the winter beauties of the forest, none are more attractive than the various forms of peziza, or cup-moulds. If a section be made through one of the cups of these beautiful fungi, they will present the appearance drawn in figure 108, plate 4. A series of hollow elongated cases will be found lying between compressed elongated tissue. In these cases a series of rather oval minute cells will be found, which are the spores of the peziza. If these are magnified with a higher power, they will be seen to be covered over with minute spines, as seen at *a*.

Amongst the objects which more especially attract the attention of observers in the country, in winter time, are the various forms of lichens, which grow parasitic upon the bark of trees. There is one of a yellow colour, which spreads on palings and the barks of trees, like dried pieces of yellow paper. At the surface of the membranous scales of which the plant is composed will be found deeper yellow spots. If one of these is cut through, and a thin section placed under the Microscope, it will be found to possess very similar organs to the peziza. A series of cases will be found, containing the minute spores by means of which the plant is reproduced. These cases, called *asci*, are figured at 109.

A walk across a damp uncultivated piece of ground will not fail to reveal some spots which are boggy. Here the bog-moss (*Sphagnum*) must be looked for, and when found, it may be regarded as a good illustration of the family of mosses, and portions preserved for microscopic examination. The leaves afford interesting examples of fibro-cellular tissue, as seen at figure 110; and this tissue may be examined from day to day, as affording an illustration of the process of development in vegetable tissue. Other forms of mosses may be found on banks, old walls, rocks, and crevices. The organs which produce the spores, or seeds, are well deserving the attention of the microscopic observer. These represent the pistils in the higher plants. The organs which represent the stamens are also very interesting, but they are not so easily procured. We therefore proceed to describe the spore-bearing organ. This may be easily seen with the naked eye, although its beauties cannot be brought fully out without the aid of the Microscope. The part which contains the spores is seated on a little stalk, and is called the "urn," and is represented in figure 112. Covering the urn, and fitting on to it like a nightcap, is the calyptra, marked *a*. On slipping off the calyptra, a conical body fitting into the urn is observed, and this is called the "operculum" (*b*). If the operculum is now lifted off, there is revealed, below, a series of twisted hair-like threads (*c*), which are called the "peristome." These processes are held together by minute teeth (*d*). The spores (*e*) are found in the interior of the urn. All these parts are subject to great varieties in different kinds of mosses.

From the mosses we may pass on to the ferns. Like the mosses, they have no regular flowers, and the parts which correspond to the urns of the mosses are the small brown scaly-looking bodies seated on the back of the fronds, or leaves. In the male fern the little brown bodies which contain the spores are round, as seen in figure 113, and in the common brakes they are placed on the edge of the fronds, as at figure 114. These organs, which are called "sori," may be easily seen as opaque objects, under the lower powers of the Microscope. In the common hart's-tongue, or scolopendrium, the sori are arranged in elongated bands. In this case the sori are covered with a membrane called an "indusium." On opening this, the sori are found lying close together. Each one of these sori is found to be made up of a number of cases called capsules, or "thecæ," attached to a stalk by which they are fixed to the frond. This organ is seen at figure 115. These thecæ are beautiful objects under the Microscope. Springing from the top of the stalk is a series of cells which surround the case, forming what is called the "annulus." This ring possesses an elastic power; so that when it breaks, the capsule is torn open, and the spores in the inside escape. The spores are covered over with little spines, as at *a*, in the same figure. The spores of ferns are often called seeds, but they are more like buds than seeds. If one of these spores is watched during its growth, it will be found that it grows into a little green membranous expansion, on the surface of which the two sets of organs resembling the pollen grains and ovules of the higher plants are developed. The representatives of the pollen grains are little moving bodies, resembling animalcules, which pass over the surface of the membranous expansion till they reach the ovules, or true spores of the fern, which they fertilize, and the young plant then shoots forth. The ferns, of which so many species may be found in a walk in the country, or cultivated in a Ward's case in town, are worthy the minute attention of the possessor of a Microscope, on account of the great variety of forms which their organs of fructification present.

The club-mosses are found on boggy moors and open places, and present a variety in the forms of their fructification. The reproductive organs are formed out of a transformed branch, and are found lying at the base of scale-like bodies, resembling the scales which form the fruit of firs and pine-trees, as seen at figure 115, a. The spores of the club-mosses are of two kinds, large and small; hence they are called "megaspores" and "microspores." The last are very minute, and when highly magnified, they present a reticulated appearance. The spores are seen at b and c in figure 117. In the interior of these spores is a minute worm-like body, which acts the part of the pollen in higher plants. The megaspores are much larger. They represent the spores of ferns, and produce an expanded membrane, on which grow the true representatives of the ovules, which coming in contact with the microspores, new plants are produced.

Another family of these flowerless plants, which has yielded highly interesting results to the microscopic observer is the group of horsetails. If these are gathered in the spring of the year, they will

PLATE 5.

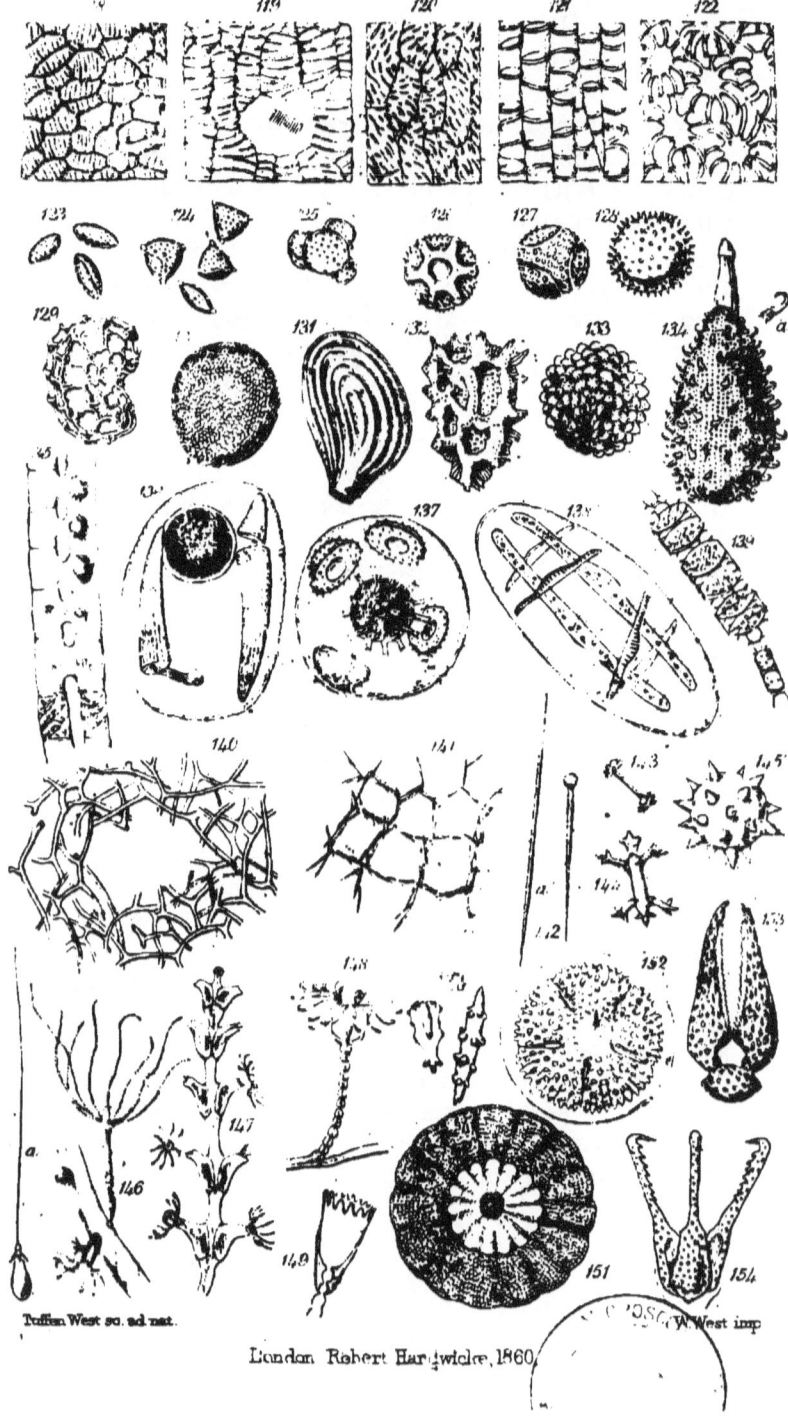

Tuffen West sc. ad nat.

London Robert Hardwicke, 1860.

W. West imp

present two forms; one showing the leaves and green parts of the fruit; the other, the leaves changed into reproductive organs. These may be very easily examined as opaque objects under the Microscope. The spores are seated on round shield-like disks, represented in plate 4, at figure 116, *a*. When the spores are examined by a higher power, they present four spiral filaments, which are twisted round the body of the spore, and seen at *b*. If the spore is breathed upon whilst under the Microscope, the spiral filaments gradually relax their grasp, and they become expanded and attached to the spore only at one end, as represented at *c*. The cuticles of the Equisetums are strongly siliceous, and are very curious and interesting objects, and will repay the trouble taken in preparing them. This may be done by boiling a piece of the stem in nitric acid and chlorate of potash, and, after washing the detached cuticle, transferring it to absolute alcohol, from thence to oil of cloves, and afterwards mounting in Canada balsam.

The study of the flowerless plants is one of never-ceasing interest. Within the last few years much has been done by the aid of the Microscope to clear away the mystery which surrounded the functions performed by certain organs they possess. Much more, however, remains to be done; and an interesting field is still open to the inquiries of the microscopist. We will now, however, take our Microscope to the pond-side, where we shall still find many plants to interest us, belonging to the lower, or flowerless groups, together with animals, the companions of their aquatic life, and the representatives of their simpler mode of existence.

CHAPTER IV.

A HALF-HOUR WITH THE MICROSCOPE AT THE POND-SIDE.

CISTERNS, ditches, ponds, and rivers, contain numerous objects to interest the microscopic observer. Some of these objects float on the surface of the water; others are found swimming about in the midst of the water; whilst the greater number are found at the bottom. In collecting objects from fresh water, little bottles may be used, and a common spoon or small net employed for collecting them. Where the objects are only few, large quantities of the water should be allowed to stand, and the whole poured off, with the exception of a table-spoonful or two, which may be then placed in a wine-glass. A little of the sediment may be taken up in a pipette or dipping-tube, and conveyed to the animalcule-cage, and the cover having been put on, it may be placed under the Microscope. If the objects are moving about too rapidly, the cover may be pressed down till they are secured. They may be first sought out with a low power, and when it is wished to examine them more closely, a higher power may be put on.

Of all the forms of microscopic plants which are found in fresh water, those belonging to the families of desmids and diatoms are most interesting. We have already spoken of plants consisting of one cell, and these also consist of one cell; but they have this peculiarity, that their cells are divided into two equal parts, each part having the same form as the other. The desmids are distinguished from the diatoms by their bright-green colour, and by their cells not depositing silex, or flinty matter, as is the case with the latter. The siliceous nature of the shells of diatoms is made apparent by their not being acted on by strong acids, as nitric and hydrochloric.

The desmids sometimes abound in ditches and small pieces of standing water. Amongst other objects in a drop of water they are easily recognized by their beautiful bilateral forms and dark-green colour. One of the most charming of these is named *Euastrum*, and consists of two notched halves of a bright-green colour, with darker green spots. It is represented at figure 28, plate 2. The green matter is composed of a waxy substance, called chlorophyle, and is the same matter as that which produces the green colour of leaves. Some of the desmids assume a lunate form, and are named *Closterium*, a species of which is figured at 29, plate 2. There are various species of *Closterium*, all of the same general form, and occasionally occurring in very great abundance. Sometimes several of the cells are attached together, forming a long chain, as in the genus *Desmidium*, seen at figure 30, from which the family takes its name. These break up and go on growing. When they grow, the new cells are formed between the two halves of the parent cells. This is represented at figures 136 and 137, plate 5. In a genus called *Scenedesmus*, several cells are united, and the two last halves are furnished with horns, as seen at figure 32; at other times several cells unite, forming a globular mass, as in *Pediastrum*, represented at figure 31. In this case each cell presents two projections, forming objects of singular beauty.

The diatoms are more numerous and widely

diffused than the desmids. The latter are decomposed, and their bodies perish when they die; but from the fact that the diatoms deposit silex in their structure, they are almost imperishable. They are found in great abundance in the mud of rivers, ponds, and lakes. They are also present in those deposits of clay which once formed the bed of rivers and lakes, and which are now dry. In order to procure the diatoms from these deposits, the clay or earth should be well washed with pure water, and the deposit allowed to subside, and the water poured off. This may be repeated several times. The deposit is then to be washed with hydrochloric acid, and when the effervescence is over, the acid is poured off, and a fresh portion is added. This may be repeated several times, and when the hydrochloric acid ceases to act, nitric acid may be employed in the same manner. When no action occurs by its use cold, the deposit may be transferred to a watch-glass, and kept over a spirit-lamp, at a temperature of about 200°, for three or four hours. The deposit must then be well washed with pure water, to remove all the acid. The deposit will be found now to consist almost entirely of diatoms. If anything else be found, it will be grains of sand. By casting the deposit into a small quantity of water, and allowing the heaviest particles alone to subside, these will be generally found to contain the sand and larger diatoms. By repeating this process successively, the deposits consist gradually of smaller and smaller diatoms, which may be examined with gradually higher powers, in proportion to their minuteness. Some are perfectly round, as in the case of the genus *Coscinodiscus*, a species of which is figured at 38, plate 2. It is marked beautifully over their surface; others are triangular; some are square, and attached together. The last form is seen in *Melosira*, species of which are figured at 36 and 37, plate 2, and 139, plate 5. The most common forms are those which are oval, or boat-shaped, and represented by species of *Pinnularia* and *Navicula* in figures 34 and 35 *a*, in plate 2. Some of these are again larger at one end than the other, as in *Surirella*, figure 33. The markings upon the surface are very various. In some forms the markings are exceedingly minute: so small are they, that certain species of diatoms have been used as test objects, for testing the highest powers of the Microscope.

Whilst living, the diatoms possess the power of moving about, and in some of them, as well as the desmids, a movement has been observed of the small particles in their interior. The diatoms are generally of a brownish or brownish-yellow colour, which seems to be due to a small quantity of iron in their composition. They are increased in the same way as the desmids, by the production of new cells between the parent frustules. This process is seen in figure 35, *a* and *b*, in plate 2. The continuance of the species in these organisms is secured by the process of conjugation and the subsequent formation of the spores. This process is exhibited in figures 135 and 136, plate 5. In some cases, however, the spore is found without the union of two cells, as in *Melosira* represented at figure 137, plate 5.

Sometimes, attached to the bottom of a pond or river, or growing from immersed objects, or floating about in the water, will be found long green filaments. These are the fronds of confervæ. All forms of these—and they are very numerous—will be found most beautiful objects for examination. They may be laid on a slip of glass in water, and covered over with a piece of thin glass; or they may be placed in the animalcule-cage. They consist of a series of cells growing end to end, and their partition-walls can be easily seen. They are of a green colour, from the chlorophyle contained in their interior. In the case of the yoke-threads, the chlorophyle is frequently arranged in a spiral manner along the interior of the filament, as in the *Zygnema* represented at figure 11, plate 1. These yoke-threads may be often seen to unite with each other, and the contents of one cell are emptied into the other, forming the spore of the plant, as seen at figure 135, plate 5. The cell contents sometimes break up into smaller portions, called zoospores, which, when they escape from the cell in which they are contained, move about with great rapidity. This is seen in figure 11, plate 1, at *a* and *b*. The moving power of the lower plants is well seen in the division of these confervæ, called *Oscillatorias*, which are sometimes found in semi-putrid water. A species is figured at 12, plate 1. As they lie upon the glass slide they will be seen to move over each other in all directions : hence their name.

Some of the spores formed by the confervæ move about by the agency of little organs called cilia. These are extensions of the motile matter of the cell, and are found very commonly in the animal kingdom. Occasionally, a number of these ciliated spores are aggregated together, forming a rapidly-moving sphere. Of this the *Pandorina Morum* affords a good example, seen at figure 13, plate 1, in which each spore possesses two cilia. But the most remarkable of this kind of moving plant is the *Volvox globator*, represented in figure 14 of the same plate. This beautiful moving plant was at one time thought to be an animalcule, but it is now regarded as a true plant. It consists of a large number of spores, or cells, each having two cilia, and connected together by a delicate network of threads. In the interior of this moving sphere are seen smaller globular masses, of a dark-green colour, which are the young of the volvox, which have not yet developed the network by means of which their spores are separated, and their ciliated ends presented to the water, and by means of which their movements are effected.

Another form which is now regarded as a locomotive plant is the *Euglena viridis*, seen at figure 15, plate 1. It is often found in prodigious numbers, giving to water the appearance of green-pea soup. When placed under the Microscope, it frequently presents a red speck, or point, at one end, and an elongated tail at the other. The red spot has been regarded as an eye; but if it is watched, it will be found the red colour will often extend from the red spot to the rest of the body; and it is probable that the red colour is only a change in the condition of the chlorophyle contained in its interior. Amongst this class of plants it is not unfrequent for the chlorophyle to assume a red colour at certain stages of its growth.

The transition from the filamentous to the membranous form of these plants is well seen in the species of *Ulva*. These are found in both fresh and sea water. In the early stages of its growth, the ulva presents the filamentous form of a conferva, as seen at *a*, in figure 26, plate 2. Gradually the cells of the filament split up into two or three seams (*b*); and this goes on till at last a broad flat membrane is produced (*c*).

If the plants of our fresh waters are interesting, not less so are the animalcules; for, just as we have one-celled plants so we have one-celled animals, and it was only by the aid of the Microscope that they were discovered and can be examined. Wherever the above plants are found, there will also be discovered animals to feed upon them. The animal is distinguished from the plant by its feeding on plants, whilst the latter feed on inorganic substances.

There is considerable difficulty in at once distinguishing between the lowest forms of animals and plants. Although the animal generally possesses a mouth, and a stomach in which to digest its vegetable food, there are some forms of animal life so simple as not to possess either of these organs. In the sediment from ponds and rivers there will frequently be found small irregular masses of living, moving matter. If these are watched, they will be found to move about and change their form constantly. As they press themselves slowly along, small portions of vegetable matter, or occasionally a diatom, mix, apparently, with their substance. Cells are produced in their interior, which bud off from the parent, and lead the same life. These creatures are called amæbas, and are represented in our first plate, figure 16. Although they have no mouth or stomach, they are referred to the animal kingdom. They appear to consist entirely of the formative matter found in the interior of all cells called moto planes or sarcode without any cell-wall. If we suppose an amœba to assume the form of a disk, and to send forth tentacles, or minute elongated processes from all sides, we should have the sun animalcule (*Actinophrys Sol*), which is represented at figure 17, plate 1. This curious creature has the power, apparently, of suddenly contracting its tentacles, and thus leaping about in the water. It can also contract its tentacles over particles of starch and animalcules, and press them into the fleshy substance in its centre. This is undoubtedly an animal, but it has no mouth or stomach. A large number of such forms present themselves under the Microscope. Some of them are covered with an external envelope, which they make artificially, by attaching small stones and other substances to their external surface, as in the case of the *Difflugiæ*, seen at figure 18, plate 1 ; or they may form a regular case, or carapace, consisting of a hairy membrane, as in *Arcella*, represented at figure 19. We shall meet again with forms resembling these when we take our Microscope to the sea-side.

One of the most common animalcules met with in fresh water, and whose presence can easily be insured by steeping a few stalks of hay in a glass of water, is the bell-shaped animalcule. These animalcules, which are called *Vorticella*, are of various sizes. Some are so large that their presence can easily be detected by the naked eye, whilst others require the highest powers of the Microscope. They are all distinguished by having a little cup-shaped body, which is placed upon a long stalk, figured at 40, in our second plate. The stalk has the peculiar power of contracting in a spiral manner, which the creature does when anything disturbs it in the slightest manner. In some species these stalks are branched, so that hundreds of these creatures are found on a single stem, forming an exceedingly beautiful object with the Microscope. The stalks of these compound vorticellæ are contracted together, so that a large mass, expanding over the whole field of the Microscope, suddenly disappears, and, "like the baseless fabric of a vision, leave not a wrack behind." A little patience, however, and the fearful creatures will once more be seen to expand themselves in all their beauty.

F

The mouth of their little cup is surrounded by cilia, which are in constant movement; and when examined minutely, they will be found to possess two apertures, through one of which currents of water pass into the body, and from the other pass out. Not unfrequently the cup breaks off its stalk. It then contracts its mouth, and proceeds to roll about free in the water. Many other curious changes in form and condition have been observed in these wonderful bell-shaped animalcules.

If, now, we go to a very dirty pond indeed, into which cesspools are emptied, and dead dogs and cats are thrown, we shall find abundant employment for our Microscope in the beautiful forms of animalcules which are placed by the Creator in these positions to clear away the dirt and filth, and prevent its destroying the life of higher animals. In such waters, amongst a host of minor forms, we are almost sure to meet with the magnificent *Paramœcium Aurelia*, figured at 39, plate 2. He moves about the water a king amongst the smaller prey, on whom he feeds without ceasing. He is of an oblong form, covered all over with cilia, and very rapid and active in his movements, as able to dart backwards as forwards, and turning round with the greatest facility. In his inside several spots are observed. If a little indigo or carmine is introduced into the water in which he lives, these spots become coloured by his taking up these substances. From this, Ehrenberg concluded that these spots were stomachs, and as such spots are very common amongst these animalcules, he called them many-stomached (*Polygastrica*). There is, however, reason to doubt the correctness of this conclusion of the great microscopist, as, although these spots exist in the body, they are not necessarily stomachs. They are, in fact, empty spaces, or vacuoles in the interior of the little fleshy lump of which the animal is composed. They are found in the vorticella, and in most of the true animalcules.

All animalcules have been called infusory, because they seem so abundant in many kinds of vegetable infusions. Ehrenberg divided them into *Polygastric* and *Rotiferous*. The last are also called wheel-animalcules, as, when looked at through the Microscope, they appear to be supplied with little wheels on the upper part of their body. The most common form of these creatures is the *Rotifer vulgaris*, represented at figure 41, plate 2. The branches or leaves of any of our common water-plants can scarcely be examined without some of those pretty little creatures being found nestling among them. The structure of these creatures is highly complicated, and the family to which it belongs is far removed from the polygastric animalcules with which it is associated by Ehrenberg. On examination, the wheels will be found to consist of two extended lobes, the edges of which are covered with cilia. These cilia are in a constant state of movement, and produce the appearance of wheels moving on an axis. Between the wheels is the entrance to the mouth, which, in many species of wheel-animalcules, is furnished with a strong pair of jaws. This leads to an œsophagus, a stomach, and an intestinal tube. Two little spots on the neck seem to indicate the existence of eyes; whilst a projecting organ, believed to be analogous to the antennæ, or feelers of insects, is seen directly below them. The tail is finished off with a pair of little nippers, by which the creature has the power of attaching itself to objects. When moving, its whole body is extended, but it has the power of drawing itself up like a telescope in its case, and appearing almost round.

The wheel-animalcules abound in our ponds and rivers, and sometimes occur in great numbers in the aquarium. The common wheel-animalcule, *Rotifer vulgaris*, is most frequently found in lead gutters and the drinking-fountains used for birds. If a little of the deposit which usually accumulates in the former is placed in a test-tube with water, and exposed to the light, in a short time the rotifers will be found swimming about in great numbers, and may be transferred to a live-box by means of a dipping-tube : if allowed to dry, they can be afterwards revived by adding a little water. Several of the wheel-animalcules are fixed, forming on the outside of their bodies a little case or tube in which they dwell. These forms are beautifully seen when illuminated by the spot lens.

CHAPTER V.

A HALF-HOUR WITH THE MICROSCOPE AT THE SEA-SIDE.

ON a visit to the sea-side, the Microscope is an essential instrument to all who would wish to study the wonders of the ocean. It is a curious fact, that the few grains of common salt in the gallon of sea-water seem to determine the existence of thousands of plants and animals. We shall therefore find living in the sea-water, plants and animals belonging to the same families as those in fresh water, but belonging to entirely different species.

The sea-weeds present strikingly different forms. Although many of them are microsopic, and belong to the families of *Diatomaceæ* and *Confervaceæ*, all the larger forms present interesting objects for examination in the structure of their fruit-bearing organs. No better subject for the latter purpose can be procured than the common bladder-wrack, which is so abundant on all our shores. If a frond of this fucus is examined, there will be found at certain parts a swollen mass, dotted over with round yellowish bodies. If one of these is taken and carefully pressed between two pieces of glass, it will present the spores surrounded with hairs of the most delicate and various structure. Some of the spores are divided into four parts, and on this account are called tetraspores. These are seen at *d*, figure 111, plate 4. The bladder-wrack is frequently covered with minute parasites; one of the most common of these is *Polysiphonia fastigiata*, which is represented at figure 111, plate 4. As seen in the drawing, this little plant is branched, and the stems present a series of flattened cells. On the branches are placed the fruit-bearing organs, in the form of little capsules, seen at *a*. These capsules contain tetraspores (*d*). At the ends of the branches are organs of another kind, representing the stamens, and which are called *antheridia*. These are seen at *e* in the same figure. The sea-weeds present a great variety in the form of these organs, and may be easily preserved for investigation in small glasses of sea-water.

The animal structures of the sea-water must now, however, claim our attention. Amongst the lowest form of animal life are the sponges. They are frequently cast on the shore with sea-weeds, and afford interesting objects for the Microscope. They are composed of animal matter, which lies upon a structure of horny, calcareous, or siliceous matter. The common sponge which is used for domestic purposes may be taken as a type of the whole group. If a thin section of the common sponge is made with a pair of sharp scissors and placed under a low power, it will be seen to be composed of a network of horny matter, represented in figure 140, plate 5. If now we take one of the common forms from our own sea-shore, we shall find that the network is composed of siliceous spicules lying one over the other, as represented in figure 141 of plate 5. If one of these spicules is examined (*a*) and compared with a spicule from another sponge, it will be found to differ in form and size; and the species of sponges can actually be made out by the shape of their spicules. Some of our British sponges have calcareous spicules. This is the case with *Grantia ciliata*. There is a little boring sponge, called

Cliona, found in the shells of old oysters, which has its spicules pin-shaped, as seen at figure 142. The fresh-water sponge has very peculiar-shaped spicula, and is represented at figure 143. In some the siliceous bodies are round, with projections, as in *Tethea*, seen in the drawing, figure 145. Sometimes the spicula assume a stellate form, and are even branched, as in the spicula of an unknown sponge given at figure 144.

Amongst the lowest forms of animal life, none are more interesting to the microscopic observer than those belonging to the family of *Foraminifera* (Hole-bearers). They are thus called on account of the minute holes which cover their shells. If we suppose a creature as simple in structure as the amœba, or sun animalcule, of which we have previously spoken, and which are figured in 16 and 17, plate 1, with the power of forming a little calcareous shell, we should have a foraminifer. Some of these shells have the form of a nautilus, and when first observed they were supposed to belong to this group of shell-fishes. In form they certainly resemble the higher forms of mollusca, as may be observed in figures 21 and 24, in plate 1. Sometimes, however, they are elongated or cone-shaped, as in figure 25. Other forms are seen in figures 20 and 22. They may often be found alive at the sea-side, nestling in the roots of the gigantic tayles which are so often thrown on the shore after a storm. If the roots of these plants (*Laminariæ*) are washed, and the deposit examined carefully, the foraminifera will be seen at the bottom of the vessel, and may be picked out one by one. When this is done, they will be found to have the power of protruding through the little holes in their shells their soft bodies, in the form of long tentacles, as seen at figure 24, in the first plate. With these they seem to have the power of moving, as well as of taking up the matters by which they are nourished. The shells of these creatures are not so small but they may be seen with the naked eye, and they need only a low power to observe all their structure. They are found at great depths in the ocean, and have been brought up by the dredge from the deepest parts of the Atlantic. They are very abundant in some rocks, especially in the chalk: they may be obtained from the latter substance by rubbing a piece of chalk with a brush in water. The water must be first decanted from the coarser particles of chalk, and in subsequent deposits the foraminifera will be found. They may be obtained from dry sand in which they are contained, by throwing the sand into water, when the sand will sink and the foraminifera will swim on the surface, and may be skimmed off. They are best examined as opaque objects.

The family of polyps will next command attention. One of the most simple forms of this family is found in ponds and rivers, and is called the fresh-water polyp or hydra. It is figured at 146, plate 5. It may be easily observed, adhering to plants, with the naked eye, and needs only a low power with transmitted light to observe it accurately. Its body is cup-shaped, surmounted with eight long tentacles, which it has the power of retracting. It produces young ones by the process of budding, and the buds may be often seen protruding from the side of their parents. It is very tenacious of life, and may be cut into several pieces, and each part will grow into a new hydra. These, with many other polyps and the jelly-fish, have their flesh filled with little hair-like bodies, which, from their property of stinging in some species, have

PLATE 6

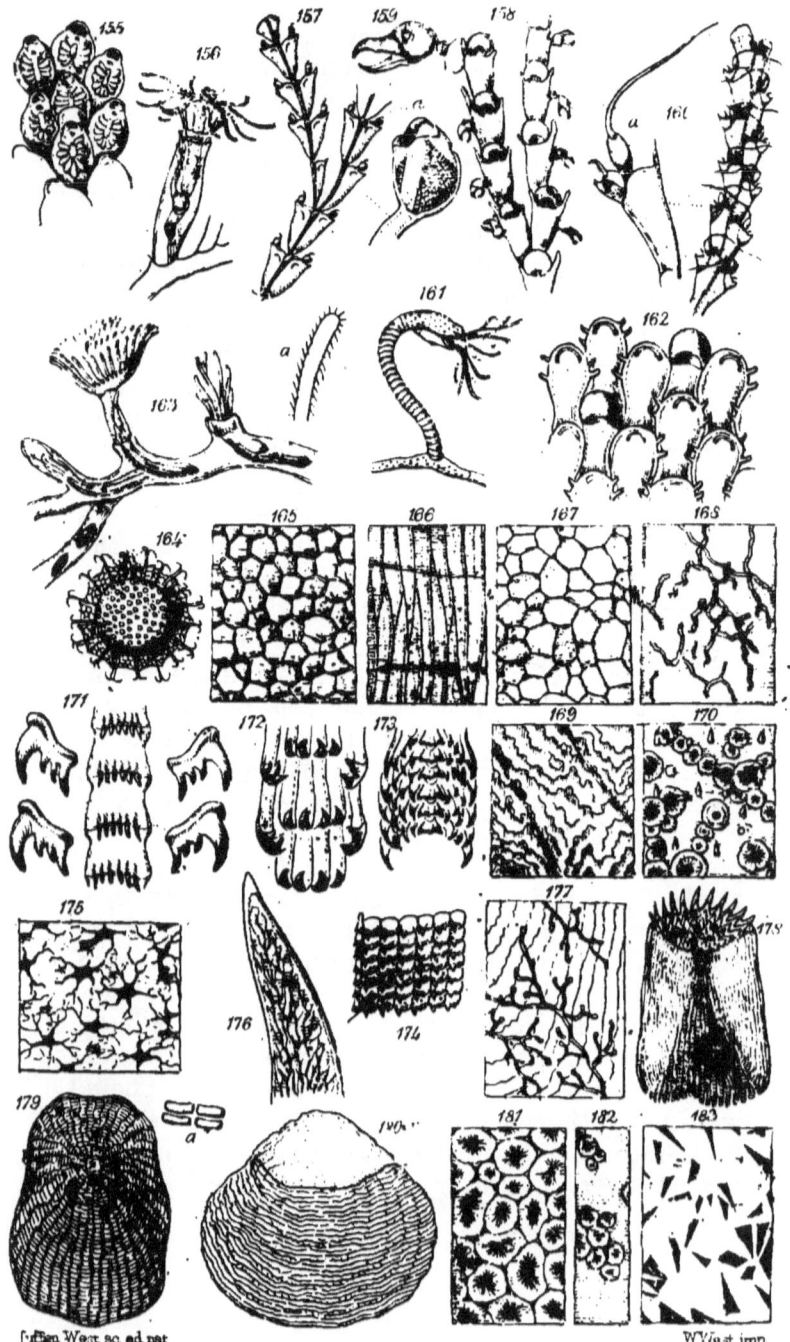

been called stinging hairs, as seen at *a*, figure 146. If we suppose several of these hydras placed in little cups upon a common branch or stem, we should have a *Sertularia*, or such an animal as is represented at Figure 147, Plate 5. These polyps are very common on all our sea-shores; and the branches and cups are often cast up on the shore, and regarded by the uninstructed as sea-weeds. The branches and cups are called the *polypidoms* of the animal, and assume a great variety of forms. When the cups are fixed on ringed stalks, they constitute the genus *Campanularia*, seen at figure 148, plate 5. These cups are often objects of great beauty, as in those of *Campanularia volubilis*, figured in 149. It is the polypidom which constitutes the coral in the family of polyps, producing the masses of carbonate of lime which sometimes cover the bottom of the ocean and form reefs in the sea. In one family of polyps, known as sea-fans (*Gorgoniæ*), which are calcareous, the fleshy mass covering the horny polypidom contains spicula of various forms, which are beautiful objects under the Microscope. These spicula are seen at figure 150, plate 5. The red coral of commerce is another interesting form of these polypidoms. In some families of these polyps, as in the campanularidæ and the coryńidæ, the young, before they arrive at their mature stage, assume the forms of minute medusæ or jelly-fishes. These are exceedingly beautiful objects for microscopic observation.

Another family of animals common enough in the sea, are the star-fishes and sea-eggs (*Echinodermata*). Although not themselves microscopic, certain parts of their structure present very interesting objects for examination. If a section is made of one of the spines of the common echinus, or sea-egg, it presents under a low power a beautifully radiated structure. This is seen at figure 151, plate 5. The suckers, also, of the same animal present little rosettes, surrounded by a very delicate hyaline disk, represented at figure 152. Upon the surfaces of both star-fishes and sea-eggs will be found little moveable bodies which are called *pedicellariæ*. In the sea-egg they possess three moveable nipper-like limbs, whilst in the common star-fish they present only two. These are represented at figures 153 and 154, plate 5. A controversy has been raised on the question as to whether these bodies are parasitic animals, or part and parcel of the structure of the creature on which they are found. As they are so constantly present, they are undoubtedly parts of the animal on which they are found. The movements of the nippers are very active, and they frequently lay hold of objects which pass near them.

As common on the shore as the polypidoms of the polyps, are the animal skeletons called, in some parts of the country, sea-mats (*Flustra foliacea*). When placed under a low power, and viewed by reflected light, the sea-mat is composed of little cavities or cells, seen at figure 162, plate 6. In each one of these is seated a creature of much more complicated organization than the polyps just examined. It has, it is true, a ring of tentacles; but if these are examined, the tentacles are found to be covered with cilia, as seen at *a*, in figure 163, plate 6. This family of creatures are called *Polyzoa*, or *Bryozoa*, and form a group of animals which are classed with the *Mollusca*, or shell-fish. Sometimes these creatures attach themselves to sea-weeds, oysters, stones, and other objects at the bottom of the sea, forming a kind of cellular membranous expansion. Such are the species of *Lepralia*, figured at 155. Sometimes the cells are elongated and elevated above the surface of the object on which they are placed, as in the case of *Bowerbankia*, seen at 156. A beautiful form of these creatures is the shepherd's-purse coral (*Notamia bursaria*), represented at figure 157. This creature belongs to a group of the polyzoa, remarkable for possessing little processes on the margins of their cells, in shape resembling the bowls of tobacco-pipes, birds' bills, and bristle-like organs. On examining them with the Microscope, they present a very complicated organization. The birds' bills possess two jaw-like processes, which open and shut like a bird's beak, and from this fact they have been called *avicularia*, or bird's-head processes (*a*). The tobacco-pipe form in *Notamia* is peculiar to that genus. In other species, as in *Bugula avicularia*, seen in figure 158, these creatures possess not only the bird's-head process, but a second, consisting of a long bristle or seta, attached by a joint to a process below (*a*). These bodies are called *vibracula*, and the bristle-like extremity is kept constantly in action, and the form of avicularia is seen in *Bugula Murrayana*, at figure 159. Both processes are seen in *Scrupularia scruposa*, at figure 160. Few objects are more curious under the Microscope than these avicularia and vibracula in a state of action. Whilst the function of the vibracula, seen at *a*, figure 160, seems to be to sweep away objects that would interfere with the life of the animal in the cell, it has been suggested by some that the avicularia secure by their jaws the food necessary for its sustenance : it seems probable, however, that they serve the purpose of a protective police. Of the various forms which the cup itself assumes, none are more interesting than those of the snake-head zoophyte, shown at figure 161, plate 6, in which it assumes the form of a snake's head, with the tentacula projecting like a many-parted tongue. The polyzoa are also inhabitants of the fresh water. Of these the most common form is the *Plumatella repens*, figured at 163. The eggs of a fresh-water species, *Cristatella mucedo*, seen in figure 164, are covered with projecting spines with double hooks at their extremities, perhaps for the purpose of catching hold of objects. Such eggs may be often found upon portions of water-lily, bulrush, and other aquatic plants which float about in our rivers, lakes, and ponds.

Although but few of the shell-fish belonging to the large class of mollusca are microscopic, yet the structure of their shells can only be investigated by the aid of the Microscope.

If any common shell be picked up on the sea-shore, it will be found to possess a rough outside, generally of a darker colour, and sometimes beautifully ornamented, whilst on the inside it is smooth, and frequently of a rose-colour. This inner smooth layer is called the *nacre* of the shell; and it is from this substance that pearls are formed in the interior of many shells. Both the outer and the inner layers present different kinds of structure in different species of shells. The outer layer can be well examined in the shell of the mollusc called the *Pinna*. The outer layer in this shell projects beyond the inner, and may be easily submitted to examination by reflected light under a low power, when it will exhibit the appearance represented at figure 166, plate 6. The external surface presents the appearance of hexagonal cellular tissue. If a portion of the shell is ground down, so as to form a very thin layer, it may be examined with transmitted light, and its hexagonal structure will be much more apparent. If a portion be examined lengthwise, it will be seen that the hexagons result from the shell being composed of a series of hexagonal prisms, as seen in the view of a longitudinal section given at figure 166, plate 6.

All bivalve shells partake, more or less, of this character; and if a portion of the outer coating of the shell of the oyster be examined, it will be found to present a general resemblance to that of the shell of the pinna, as seen at figure 167. In many shells the inner layer is almost structureless, but in those cases where the smooth white appearance is presented which is called *mother-of-pearl*, it consists of a series of waved laminæ lying irregularly one on the top of the other; represented at figure 169. In other shells this membranous internal layer is traversed by minute tubes, as is seen in the genus *Anomia*, seen at figure 168. This structure has been considered due to the natural form of the shell; but late investigations lead to the conclusion that these tubules are the borings of some parasitic animal.

The shells of the crustacea also present a series of very interesting structural differences. The shell of the common prawn, when mounted in Canada balsam, or examined in water or glycerine, presents a series of bodies looking like nucleated cells. These are seen in figure 170, plate 6. Many shells present this appearance, and it was at one time supposed to indicate clearly that the shell originates in cell-growth as well as other parts of the structure of an animal. It has been, however, recently shown, that such appearances as that presented by the prawn-shell may be produced by the crystallization of inorganic salts in contact with organic substances in solution, independent of a living organism.

Surprising as it may seem to some persons, the teeth of mollusca afford beautiful objects for microscopic examination. All that is necessary to examine these organs is, to take the palate, or tongue, as it is called, of any of our common molluscs, and to stretch it on a glass slide, when it may be seen by transmitted or reflected light. In the common whelk, the teeth are placed in rows, and are composed of a broad base with four projecting points, the two outer of which are larger than the inner, as seen in figure 171, plate 6. In the limpet, the teeth present four projections, which are all of the same size; seen in figure 172. In the common periwinkle another kind of arrangement is observed, and is figured at 173.

When sea-side specimens have been observed and put up, the fresh-water mollusca may be next investigated. Here other forms will be observed. The species of the genus *Limneus* are found in every pond, and kept in every aquarium. The tongues of these creatures, represented at figure 174, will give a lively idea of the nature of the scavengering processes they carry on.

The scales of fishes are interesting microscopic objects. The structure of these organs indicates the family of fishes to which they belong. It is in this way that a single scale found in a rock will throw a light on the nature of the fishes which inhabited the seas or rivers from which the rock was deposited.

Fishes' scales have been called *ganoid, placoid, cycloid,* and *ctenoid,* according to the families to which they belong. The sturgeon has *ganoid* scales. They are shiny, and have a structure like bone, and are represented at figure 175, plate 6.

The sharks, rays, and skates have *placoid* scales. They are frequently terminated with a prickle, as in the scales of the skate; seen at figure 176.

This structure resembles the tubular structure in the teeth of the higher animals.

Fish-scales are frequently permeated with minute tubes, drawn in figure 177, plate 6. These appear to be the work of some minute parasite, such as that producing the tubules in shells, and which has hitherto evaded the scrutinizing investigation of the microscopic observer.

The fishes of the earlier rocks belong to the *ganoid* and *placoid* groups. The great majority of recent fishes belong to the remaining groups. The common sole affords an instance of the *ctenoid*, or comb-like scale, seen at figure 178.

The *cycloid*, or circular scales, are found in such fish as the whiting, and represented at figure 179. It is not uncommon to find in these scales calcareous particles, shown at *a*. In the sprat the *cycloid* scale assumes a form almost as broad as it is long, and is seen in figure 180.

The examination of these hard structures in the marine creatures is a good preparation for the further study of those hard parts in the higher animals to which the name of bone and ivory is given. Such things may, however, be procured in the house; and when the rain is falling, the seaside forsaken, or the country miserable-looking, we can still enjoy the long winter evenings with our Microscope in the house.

CHAPTER VI.

A HALF-HOUR WITH THE MICROSCOPE IN-DOORS.

For amusement and instruction with the Microscope, we need scarcely stir out of our rooms. The very hairs on our head may be made objects of interesting investigation, and especially if we compare them with the hairs of other animals, and the appendages generally of the skin. The fine outer coating of the skin is composed of minute scales, which are flattened cells, and may be easily observed by scraping a portion of the skin on to a glass slide with a drop of water on it. The nails, the hairs, and other appendages of the skin, are composed of the same kind of scales, or cells. These cells are developed in little pits, or follicles, from which the hair is projected, as it were, by their growth from below. Under a low power the cells of the human hair cannot be observed. It presents, however, a well-marked distinction between the outside, or *cortical layer,* and the interior, or *pulp.* The latter, by a high power, especially if the hair has been first submitted to the action of sulphuric acid, will be found to contain cells more or less spherical, whilst the former contains cells more or less flattened. These project a little beyond the edge of the hair, so that its sides are not quite smooth, as seen at figure 184 in plate 7. By placing a hair between two pieces of cork, fine transverse sections of it may be made by means of a sharp razor. If these are put under the Micro-

PLATE 7

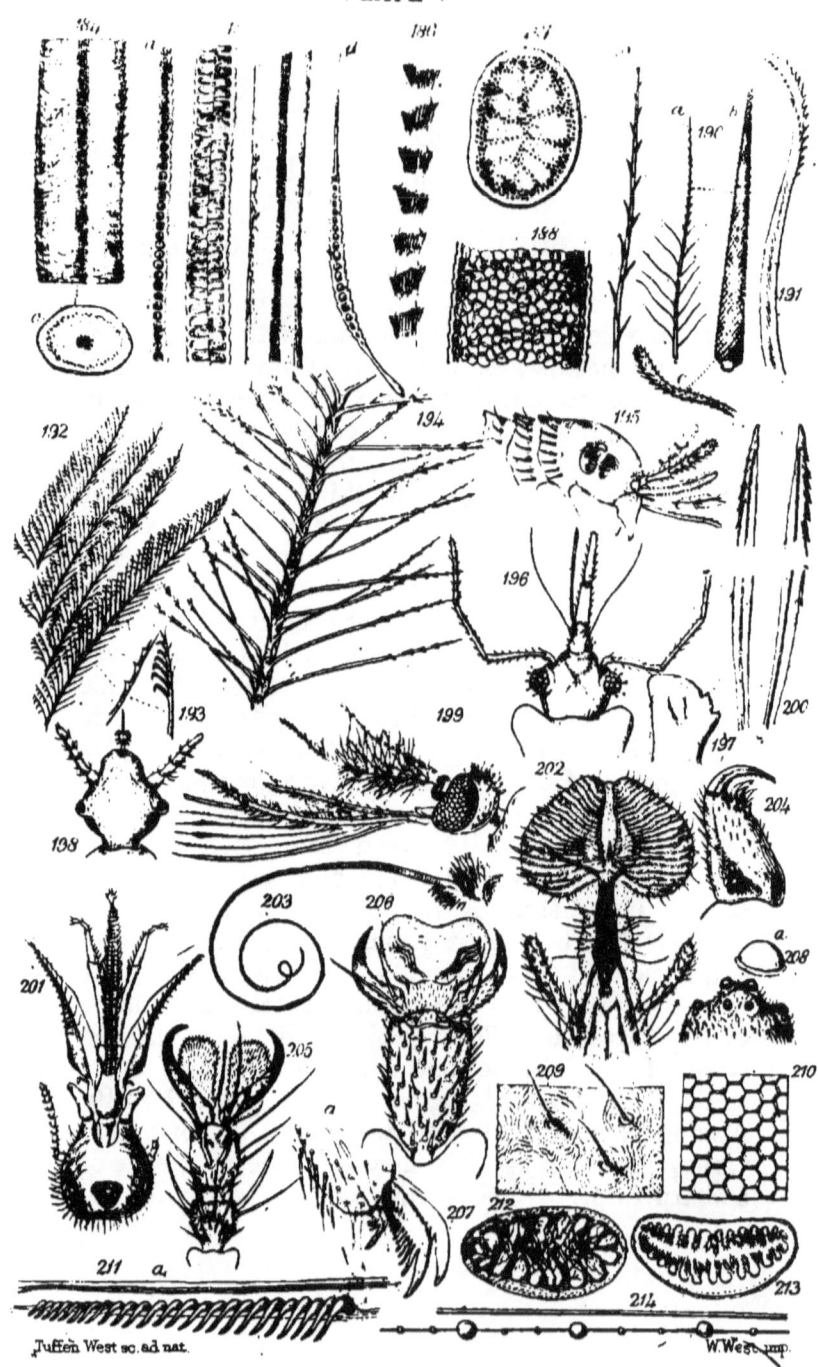

Tuffen West sc. ad nat.   W. West imp.

London. Robert Hardwicke, 1860.

scope, the pulpy portions will present a dark appearance in the centre, as seen at *a*. The hairs of animals offer a great variety in the disposition of the cells of which they are composed. The hairs of the mouse present a series of dark partitions running across the hair between the cells. In the younger hairs, these partitions are single, as represented at *a* in figure 185, plate 7; whilst in the older ones they appear double, as seen at *b*. The hairs from the ear of the mouse present these dark partitions very distinctly, shown at *d*. Such hairs stand intermediate between true hair, a section of which is seen at *c*, and wool. A piece of flannel or blanket will afford a good illustration of the latter. This is figured at 235 in the 8th plate. In this case it will be seen that the scales, or cells, of the cortical part, project beyond the surface, and render the wool rough. This roughness of the outside is supposed to render such hairs fitted to be used in the process of felting; the rough sides of the hairs adhering together. The chemical composition of the hair has also something to do with this process. Human and other smooth hairs, will not felt.

The fibres of plants used in weaving may be conveniently compared with hairs derived from the animal kingdom. The woody fibre of the flax may be obtained from a linen handkerchief. A linen fibre is represented at *b* in figure 234, plate 8. The apparent knots in the fibre arise from injury in the uses to which the fabric has been applied. The original fibres have no such fractures, as shown at *a*, and are perfectly smooth. So are the fibres of silk, represented at figure 236. Cotton-wool is produced from the inner surface of the pod, or fruit of the cotton-plant, and is figured at figure 237. It becomes twisted during its growth, and although not so strong as linen or silk, its irregular surfaces permit its being spun into a strong yarn, from which all cotton fabrics are made. The young microscopist should make himself acquainted with the forms of these various fibres; as, from their being so constantly present in rooms where the Microscope is used, and occasionally employed in cleaning the apparatus, they often present themselves as foreign substances, among other objects that are being examined.

It is also interesting, and sometimes of importance, to be able to ascertain of what material a fabric may be composed. Thus by means of the Microscope, and that alone, we know certainly that the cere-cloths in which Egyptian mummies are wrapped is a linen fabric, whilst the similar investment of Peruvian mummies is cotton. The hair of the bat, represented at figure 186, plate 7, presents a singular instance of the projection of the scales, or cells, in a regular form. Hairs are not often perfectly round;—in the peccary they are oval, as seen in figure 187, plate 7. If a transverse section of this hair is examined, it will be found that the cortical substance projects completely into the pulpy part of the hair in several places, so as to break up the pulp into several separate sections.

In some cases it is not easy to distinguish between outside and inside structure, as seen in the hair of the musk-deer, in which the whole is found to consist of a mass of hexagonal cellular tissue, similar to that seen in the pith of plants. This hair is shown in plate 7, figure 188.

Insects are frequently covered with hairs, especially in their larva, or caterpillar state. These hairs when stiff and sharp, penetrate the skin, and produce irritation there. This is the case with the large tiger caterpillar. The hairs of this caterpillar are furnished with a series of barbs, which, when they once penetrate the skin, are not easily removed, as seen in figure 189, plate 7.

Spiders are frequently covered with hairs, some of which are branched, as at *a* in figure 190; others present a spiral appearance, seen at *b*; whilst, again, others offer a series of small bristle-like hairs running down each side of the primitive hair, which will be seen at *c*.

Many of the crustacea have hairs upon their shells. Those upon the flabellum of the common crab have minute bristles on one side of the parent stalk, so as to form a little comb, with which to brush off the impurities from its branchiæ. This structure is seen at figure 191 in plate 7. A live crab from the aquarium may be watched for the purpose of observing these cleanly movements.

The study of the uses of the epidermal appendages is one full of interest, as in no one set of structures do we find a greater variety of adaptations of a common plan to the wants of the creatures in which they are found. The feathers of birds belong to the same type of structure as the hairs of animals. If the pinnæ of a common goose-quill, used for a pen, are examined, the pinnules will be found to be covered with minute hooks, drawn in figures 192 and 193, plate 7. These hooks on the upper surface are so arranged that they catch the nearly plain and slightly toothed pinnules on the lower side.

The down from the feathers of the swan, with which pillows and beds are stuffed, is also a beautiful object, and its microscopic structure will at once reveal the cause of its lightness, softness, and warmth. This is seen at figure 194, in the 7th plate.

Amongst the creatures which domesticate with

us are certain insects which are more frequently discovered than acknowledged. However disagreeable their presence may be, they become interesting objects for microscopic investigations, and are not less calculated to excite our admiration than creatures more ceremoniously treated. We first call attention to the common flea (*Pulex irritans*). This beautiful insect belongs to a large family, each species of which has its peculiar habitat in the epidermal appendages of some of the higher animals. The head of the human flea may be taken as the type of the family. This is represented with great accuracy at figure 195, in plate 7. It is furnished with antennæ, mandibles, and a pair of lancet-shaped jaws, with which it makes little wounds in the skin, and into which it pours the irritating secretion which renders its bite a source of annoyance. Its eye, large hind legs, and ornamental saddle on its back, are all deserving of attention.

Let us now seek another too common inhabitant of London houses, the bed-bug (*Cimex lectularius*), and, having decapitated him, submit his head to a low power. He, too, is a biting creature; and you will observe, as drawn in figure 196, that his jaws are finer than those of the flea, and are like a pair of excessively fine sharp hairs; they are inclosed in a sheath, from whence they are projected when used. In the same sheath is the tongue, which performs the double office of depositing in the wound an acrid and irritating secretion and sucking up the blood of its victim. The antennæ and eyes of the bug are also worthy of examination. From the latter will be found projecting minute hairs.

A still more despised animal may now be sought (*Pediculus*). It also belongs to a large family, and each mammal and bird seems to be attended with its peculiar louse. Two species are found in dirty and diseased conditions of the human body. Disgusting as connected with want of cleanliness, they are, nevertheless, perfectly harmless. The head and mouth, drawn in figure 198, indicate that these creatures are adapted to live on the secretions of the skin. The above animals all belong to the much larger group of creatures adapted to live as parasites upon other animals.

The head of the common gnat, figured at 199, in plate 7, may be now examined for the sake of comparison. In this creature, the eye of the insect may be studied. It is what is called a *compound* eye, and is composed of innumerable small lenses; each one of which is connected with a twig of the optic nerve, and capable of receiving impressions from external objects. The little lenses terminate on the convex surface of the eye, presenting an immense number of hexagonal facets. These are seen at figure 210, plate 7. In the common house-fly, there are said to be 4,000 of these facets; and in the cabbage-butterfly 17,000. The antennæ of the gnat are very beautiful; and, in fact, these organs in insects afford an endless variety of forms. At their base, in the gnat, is seen a round process on which these are seated, and it has been supposed that they are organs of hearing. Whether they are organs of hearing or not, it is very certain that they are organs of touch, and the creature is very susceptible of the slightest stimulus applied to them.

The head of the honey-bee may be now examined; and if a careful dissection is made of its mouth, a marvellous apparatus is unfolded to view, which is exhibited in figure 201, plate 7. At the base is seated the so-called *mentum*, and on each side are placed the *mandibles;* above these, and longer, are the *maxillæ,* and on each side of the prolonged central organ, called the tongue, are placed the labial palpi. The tongue can be retracted between the palpi as into a sheath. It is marked by a series of annular divisions, and, by a high power, will be seen to be covered over with hairs. This is the organ by means of which the bee " gathers honey all the day."

Whilst examining the bee, its sting may be taken out and placed under a low power, when it will be found to present the appearance of a pair of spears set with recurved barbs, which run part of the way down one side of each half of the sting. This is seen in the 7th plate, figure 200. Each of these spears is grooved on the opposite side, the two, when united, forming a canal, down which are poured the contents of the poison-bag, producing the painful effects of wounds from these instruments.

To return to the head and mouth of insects :— The tongue of the bee may now be compared with the same organ in the butterflies, which in them assumes the form of a proboscis, and is called the *haustellum,* seen at figure 203, plate 7. This instrument is coiled up when the insect is at rest, and is the organ by means of which the creature sucks up its nutriment from the flower. It has a series of lines running across it.

If the head of the common blowfly be now examined, it will be seen that the tongue, instead of being elongated as in the latter instances, is expanded laterally. This is represented in figure 202, plate 7. It is a very beautiful object, and when viewed by transmitted light, a series of spiral bands are observed to wind across each half of the tongue.

The head of the common garden spider (*Eperia diadema*) presents an interesting development of the mandibles. These organs are in pairs; each mandible consists of two joints: one is small, sharp, and hooked; whilst the other is large and short, and contains within it a bag, or poison-gland; so that when the creature seizes its prey, the bag is pressed on, and a drop of the poison exudes. This organ is represented in figure 204, plate 7. This structure is similar to what is met with in the poisonous serpents, where a poison-bag is seated at the base of a tubular tooth.

The description above given is the generally received one; but Mr. John Blackwall, our greatest authority on spiders, considers the use of the term "mandibles" to parts entirely without the mouth objectionable; he has accordingly bestowed the name of "falces" upon them. Some carefully-conducted and interesting experiments of his on their so-called poisonous secretion seem to throw great doubts on the propriety of regarding them in this light, and he has been led to consider that the purposes of it may rather be to deaden pain and still the struggles of a captured animal, as chloroform is given previous to and during operations on human beings.

The head of the spider affords also a good example of what are called *simple* eyes. Besides the compound ones before mentioned, insects have also these simple eyes—drawn at figure 208, plate 7. They consist of a single lens, as seen at *a*, and are placed in various positions in the heads of spiders.

The skin of the common garden spider is covered with hairs. These appear to surmount a series of concentric plates, seen at figure 209, plate 7. They vary in form in different species of spider;

and the skin of all should be examined for the purpose of observing these differences. The web of the spider should also be examined. The cords of these beautiful structures, which run from the centre to the circumference of the web, are plain, as seen at figure 214 ; whilst those which form the concentric lines are beaded with drops of a glutinous substance. It is by means of this adhesive matter that the webs are held together. Nor should the microscopist neglect examining the spinnarets of the spider, by which these beautiful threads are elaborated.

The breathing organs of insects are well deserving attention. Their bodies are perforated at the sides, and the openings thus formed, called *spiracles*, lead into tubes which are branched, and are called *tracheæ*. These air-tubes are composed of a delicate membrane, which is supported on a series of delicate rings, which are easily traced into the more minute branches. They are well seen in the larvæ of most of the lepidopterous insects, and represented from a caterpillar in figure 222, plate 8. The spiracle is not an open hole. In the common house-fly, seen at figure 212, plate 7, and the water-beetle (*Dyticus*), in figure 213, it is covered over with irregular branched processes from the sides of the opening. The object of this obstruction is probably to prevent particles of dust, and other foreign substances, from entering the air-passages, and thus choking the animal.

The legs of insects will afford an almost unlimited supply of objects for examination. The spoilt specimens of a summer's capture may well supply materials for a winter's examination. The legs of insects are composed generally of five parts, jointed together. The lowest of these is called the *tarsus*, or foot. It is variously formed to adapt it

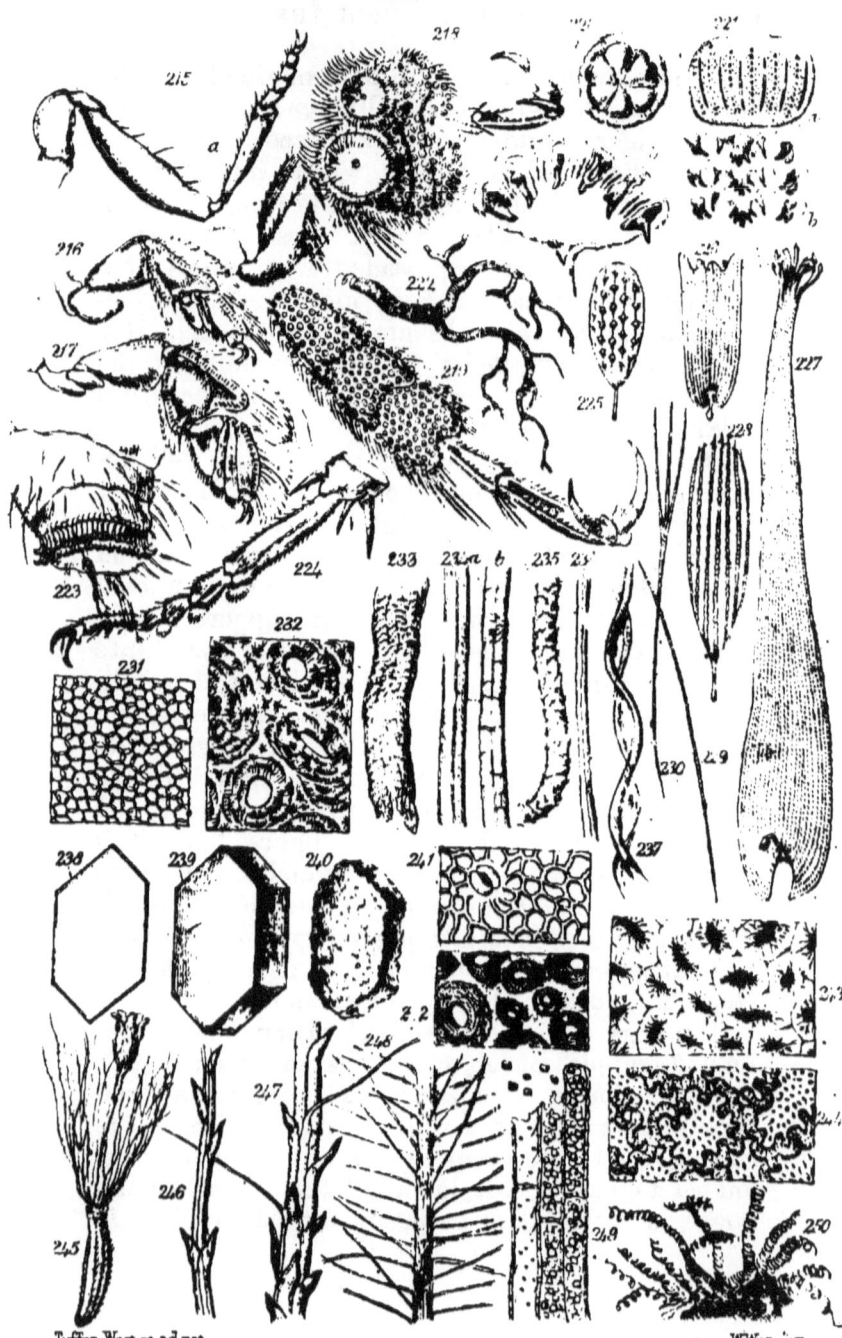

PLATE 8

London: Robert Hardwicke, 1860.

to the locomotive habits of the insect. In the common fly it is terminated with a pair of disks, which are covered with suckers, called *pulvilli*. Those of the *Empis*, a species of fly, are drawn at figure 205, plate 7. By means of these suckers the animal is enabled to lay hold of smooth surfaces, and thus to crawl up them. They also exude a glutinous matter, which assists in this process. The same kind of arrangement is observed in the common bee, represented in figure 206. The feet are also covered with hairs, and are frequently supplied with hooked joints, which assist the animals in laying hold of rough objects where their suckers would be of no use. In the spider there are no suckers, but the hooked joints and hairs enable the creature to crawl with facility. These hooks are seen in the foot of the spider in figure 207, plate 7. In the *Dyticus* the fore leg is supplied with two large suckers, which are seen in figure 218, plate 8, besides a number of smaller ones, and a hook; whilst the foot of the middle leg is destitute of the large suckers, as seen at figure 219.

The legs of beetles are often covered with little cushion-like bodies, which undoubtedly act as suckers. These are seen at figures 215, 216, 217. The three legs often differ very much from each other, and probably perform modified functions, according to their structure. This is well seen in the legs of the whirligig-beetle (*Gyrinus natator*), in which the first leg, in figure 215, is very much elongated, whilst the third is broad and short, as at figure 217, and adapted for swimming, from its oar-like form. The second leg, seen at figure 216, is intermediate in form and size.

As a contrast to these legs, adapted for the varied functions of the perfect insect, the leg of any common caterpillar may be examined; when it will be found to consist, at its extremity, of a little sac surmounted with hooks. These hooks are represented in figure 223, plate 8.

The wings of insects, too, are beautiful objects; easily investigated by a low power. The nerves which run through them are supplied with tracheæ, and they thus become organs of respiration. The under wing of the bee is supplied with a series of hooks, seen at figure 211, plate 7, which slide on a thickened nerve on the upper wing, marked *a*, and keep the wings steady during flight.

The *lepidopterous* insects, including the butterflies and moths, have got their name from the scales on their wings. These scales assume a wonderful variety of form, and claim a large amount of attention from the microscopic observer, and cannot be neglected by the entomologist.

The little blue argus butterfly has scales in the shape of a battledore, drawn at figure 225, plate 8, the handle being the part attached to the wing. All the scales have handles of this sort, whatever be their shape. At figure 226, a scale of ordinary shape is represented. Sometimes the scale is broad at the base, and pointed at top. In the meadow-brown butterfly, the point is surmounted with little clubbed projections, drawn at figure 227. Scales are found on other insects besides moths and butterflies: thus they are found on the common gnat. These are shown at figure 228. Besides their curious forms, the scales are marked with lines which are exceedingly delicate, and require the highest powers of the Microscope to bring them out. Some of the scales are thus used as tests for the powers of the Microscope.

Just as we have seen in the tongues and legs of insects, the same parts expanded or compressed according to the wants of the animal, so we find the scales assuming various forms. The scales stand in exactly the same relation to the hairs in insects, that the scales of fishes and reptiles do to the feathers of birds and the hairs of mammals. Hair-like scales are therefore not uncommon. At figures 229 and 230, such scales are represented, and may be found on the common clothes-moth.

The young microscopist, for whom our book is written, and with which we hope to make him dissatisfied, in order to facilitate his progress in natural history inquiries, will not spend much time in making dissections. Should he wish to do so, he well find the structure of insects full of interest. He has only to open a cockroach to see how curiously their digestive apparatus is constructed. This insect has a gizzard, and at the upper part it is beset with six conical teeth, as seen at *a*, in figure 220, plate 8; these teeth, working together, reduce its food to a pultaceous mass previous to digestion. When cut open, their position and relations can be easily seen, as figured at *b*. The gizzard of the cricket is also supplied with teeth, seen at *a*, figure 221; it has three longitudinal series of teeth, and each row in each series contains seven teeth. The family of insects to which the cricket belongs (*Orthoptera*) affords several other instances of the same kind of structure in the gizzard. It will be interesting to compare these teeth of the insects with those of the mollusca and the wheel animalcules.

We must satisfy ourselves with having shown the student the way to cultivate a large field of interesting and instructive phenomena in the insect world, without going further into detail.

The tissues or textures of which animals are built up or made may be easily procured in-doors. We have spoken of the hard parts which form the outer skeleton of the lower animals, as the molluscs, crabs, and fishes; the internal skeleton of the higher animals affords a not less interesting field of research. If we take a piece of bone, and having ground it so fine that we may examine it with transmitted light under the Microscope, we shall find it composed of a number of minute insect-shaped cells, surrounding an open canal, as seen at figure 232, plate 8. These cells, which are called *lacunæ*, and their little branches *canaliculi*, are modifications of the cells found in fishes' scales, and figured at 175, plate 6.

These curiously-shaped cells differ in size and form in the various classes of animals belonging to the sub-kingdom *Vertebrata*, and thus a small portion of a bone will frequently serve to indicate whether an animal belonged to fishes, reptiles, birds, or mammals. This is a matter of importance to the geologist in determining the character of the inhabitants of the earth at former periods of its history. A section of whalebone is figured at 242, plate 8.

The shells of eggs seems to be formed on the same general principles as other hard parts, and the tendency to the formation of cells with *canaliculi* may be easily observed, as in the section of a common egg-shell, represented at figure 181, plate 6. The young egg-shell should be examined, a section of which is seen at 182, if the object is to study the history of the development of the shell; and this may be compared on the one hand with the shells of the *Mollusca* and the *Crustacea*, and on the other hand with those of the scales, teeth, and bones of the vertebrate animals. Egg-shells present very different appearances. The shell of the emu, for instance, exhibits a series of dark triangular spots, and is represented at figure 183, plate 6.

As one of the hard parts of animals, the structure of cartilage is very interesting. A slice may be obtained from the gristle of any young animal. Its structure is best seen in the mouse's ear, represented at figure 231, plate 8. No one who looks at this object can but be struck with its resemblance to vegetable tissue; and it was this resemblance which led to the application of the cell theory of development, which had been made out in vegetable structures, to those of animals.

Many of the soft parts of animal tissues afford instructive objects under the Microscope. If the tongue is scraped, and a drop of the saliva thus procured placed under the Microscope, it will be found to contain many flat, irregular, scale-like bodies with a nucleus in the centre, such as are seen at figure 4, plate 1. These scales are flattened cells, and closely resemble those found on the surface of the skin. Cells of a different kind line the air-passages. If a snip be taken from inside the nostril of a recently killed ox or sheep, it will be found to be composed of cells which are fringed with cilia at the top. These are seen at Figure 5, Plate 1. These cilia are constantly moving, and produce the motion of the mucus on the surface of these passages which is essential to their healthy action.

The blood of animals presents us with objects of high interest. The human blood consists of a liquid in which float two kinds of cells. They are discoid bodies, from the three-thousandth to the three-thousand-five-hundredth of an inch in diameter ($\frac{1}{3000}$ to $\frac{1}{3500}$), and about a fourth of that size in thickness. They are represented at figure 6, plate 1. They are of two sorts—pale and red; the latter are rather smaller, but are by far the most abundant. They present a little spot in the centre, which is called a *nucleus,* and this again another little spot, which is called a *nucleolus.* The red globules vary much in size and form in different animals. Thus, in birds, reptiles, and fishes, they are oval instead of round ; and, mostly, in these three classes much larger than in mammals. This is especially the case in the *batrachian* reptiles, to which the frog and toad belong. Those from the frog are shown at figure 8, plate 1. In the fowl, shown at figure 7, and in the sole, seen at figure 9, they are nearly twice as large as in man. In the insects they are also frequently of large size, as in the cockchafer, seen at figure 10.

The proof that blood-stains have been produced by human blood on articles of dress and other things, is frequently important in medico-legal investigations. Although it cannot be distinguished from all other kinds of blood, it may be from some ; and the Microscope has been employed as an adjunct in such cases.

The structure of the skin, and other organs of the body, are very interesting subjects for microscopical investigation ; and volumes have been written upon their diversified details. The structure of voluntary and involuntary muscular tissue may be easily examined, especially the former, by taking a portion of the flesh of any animal usually eaten as food. The striated fibrillæ of voluntary muscle may be best seen in flesh cooked as food. A muscle consists of bundles of fibres, and each of these fibres consists of several fibrillæ lying close together. Each of these fibrils is seen to be crossed with lines, represented in figure 233, plate 8. These lines indicate the point of union of the string of cells which form the ultimate parts of the muscular tissue.

The structure of nervous tissue is also one of

high interest to the physiologist, but it requires the highest powers of the Microscope, and great skill in manipulation, to make out.

CHAPTER VII.

A HALF-HOUR WITH POLARIZED LIGHT.

WHAT is polarized light, and in what does it differ from ordinary light? This question is often asked, and, like many other questions in physical and natural science, more easily asked than answered.

To enable the young microscopist to form some conception of the difference between common or ordinary light and that known as polarized light, it will be necessary to form some definite idea of light itself.

Light, according to the modern theory, is produced by the vibrations or undulations of an imaginary fluid called ether; this is supposed to be a rare and highly elastic fluid, occupying all space and pervading all bodies: the vibrations of this medium produce light, just as the vibrations of air produce sound.

The length of these vibrations is inconceivably minute, and their rapidity is represented by numbers which the human mind can scarcely comprehend. Upon the relative lengths of these vibrations depend the differences of colour, red being produced by the longest, and violet by the shortest waves or vibrations.

For the production of the red ray, 37,640, and for the violet ray 59,750 undulations in an inch are requisite. In the production of the red ray 458 millions of millions, and in the violet ray 727 millions of millions of undulations take place in a second of time. White light is produced when the undulations are 44,440 in an inch, and their number in a second of time amounts to 541 millions of millions. These vibrations are communicated to the retina and optic nerve, and from thence to the brain. The rapidity with which these undulations are communicated from their source to the eye may be imagined when it is stated that the light from the sun (a distance of about 90 millions of miles) reaches us in 8 minutes and 13 seconds; a railway train travelling at the speed of 60 miles an hour would require 180 years to accomplish the same distance. The light from remotest nebula (according to Sir W. Herschel) would, however, require 2,000,000 years to reach the earth.

A ray of common light is supposed to have at least *two* sets of vibrations; viz., one vertical (or up and down), and the other horizontal (or from side to side).

These vibrations are capable of being separated either by reflection or by passing the ray through certain transparent substances. The light is then said to be *polarized*. The name is not, perhaps, the best that could have been chosen, but as it has been in use for many years, any alteration would be attended with inconvenience.

The terms poles and polarity are usually employed to describe the contrary properties possessed by the opposite ends of bodies. Thus, we have the north and south poles of a magnet, one of which attracts what the other repels; and when it was found that the sides of a beam of light, when reflected or transmitted under certain conditions, possessed opposite properties, the ray was said to be polarized from a fancied resemblance to the poles of a magnet or galvanic battery.

H

An imaginary section of a beam of common light is usually represented thus : —— —|—— and, of a beam of polarized light. ————— or In the following diagrams we shall represent the ordinary beam by three, the ordinary polarized ray by two parallel lines, and the extraordinary polarized ray by a single line.

≡≡≡≡≡≡≡ ≡≡≡≡≡ ————

If a ray of light (Fig. 16) *b* impinges on a bundle of glass plates, *a*, placed at the polarizing angle of glass (56° 45') the ray is in part reflected and in part transmitted, and both become polarized; *c* is termed the ordinary, and *d* the extraordinary ray.

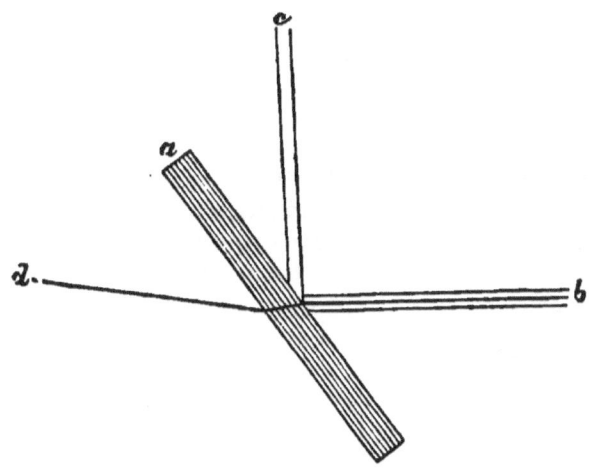

Fig. 16.

a, bundle of plates of thin glass; *b*, ray of ordinary light; *c*, ray polarized by reflection; *d*, ray polarized by refraction.

A polarized ray may be obtained by reflection

from most polished surfaces, such as a mahogany table, a tea-tray, a piece of japanned leather, &c.

During the earlier and later periods of the day, the light reflected from that portion of the sky opposite the sun is always polarized.

It will thus be seen that polarized light is of common occurrence, but the unassisted eye is unable to detect it, although one-half of the ordinary beam is lost. We may here remark that the loss of light caused by various optical contrivances is not usually detected by the eye. This is well illustrated by the Binocular Microscope. Let an object be examined with the tube directly over the prism with the prism in position; if we remove the eye for an instant, and withdraw the prism, no difference will be detected, although in the latter case double the amount of light has been transmitted through the tube.

This non-appreciation of an increase or diminution of light to the extent of 50 per cent. is perhaps owing to the dilation and contraction of the pupil of the eye.

If the reflected and refracted beams of polarized light are thrown simultaneously on a white ceiling and a white screen, the spectator will observe two spots of light of equal intensity. A polarized ray may be obtained—

   1. By reflection.
   2. ,, simple refraction.
   3. ,, double refraction.
   4. ,, transmission through a plate of tourmaline or crystal of herapathite.

The diagram (fig. 16) on page 98 represents the two first, c being the reflected, and d the refracted ray.

The polarization of a ray of ordinary light by

double refraction is shown in fig. 17; *a* is a rhomboidal crystal of Iceland spar. These crystals have

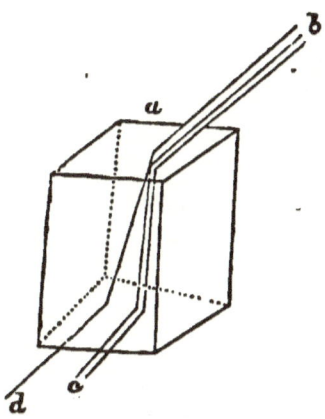

Fig. 17.

a, rhomb of Iceland spar; *b*, ray of common light; *c*, ordinary ray of polarized light; *d*, extraordinary ray of polarized light.

the property of splitting the impinging ray into two; thus, if a small hole is made in a card, and viewed through a rhomb of Iceland spar, two discs of light will be seen; or, if a black line is drawn on a piece of paper, two images of it will appear.

b is the ray of common light which becomes divided as it passes through the crystal. These rays are both polarized.

Certain varieties of tourmaline, when cut into plates parallel to the axis of the crystal, possess the property of polarizing common light. Fig. 18 represents such a plate.

Having seen how a polarized ray can be obtained, the reader will ask, How am I to recognize this condition of light; for you have already told me that it is not to be detected by the unassisted eye?

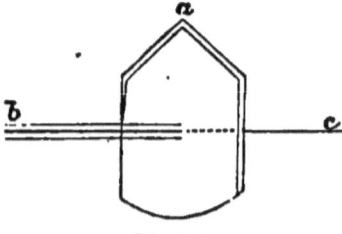

Fig. 18.

a, plate of tourmaline; *b*, ray of common light; *c*, ray of polarized light.

In order to distinguish the difference between ordinary light and that which has become polarized, special means are required for that purpose. It is an axiom that the medium capable of producing polarized light is also capable of analyzing it. Thus, if the reflected ray *c* (Fig. 16, page 96) is reflected on a mirror whose surface coincides with that of the polarizer, the ray will be reflected in the same manner as an ordinary ray; but if we gradually revolve it until it stands at right angles

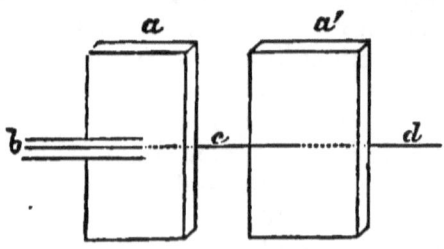

Fig. 19.

a a', two slices of tourmaline with angles coincident; *b*, beam of common light; *c*, polarized ray; *d*, ditto transmitted.

to the polarized, the ray is intercepted and destroyed.

Let a a' represent two plates of tourmaline with their angles coincident, a is the polarizer and a' the analyzer; with the plates in this position, the polarized ray c passes through to d (Fig. 19).

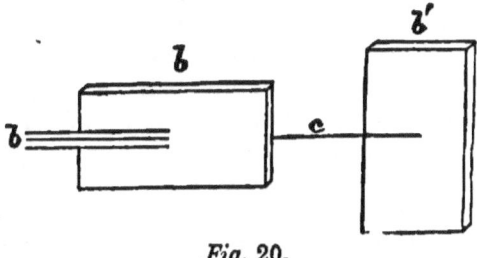

Fig. 20.

b b', two slices of tourmaline crossed; b, beam of common light; c, polarized ray stopped by b'.

If we now cross the plates, the ray c is no longer transmitted. If the analyzer is now revolved another 90°, the ray is again transmitted. Revolve it 90° more, the ray is stopped; and, on the completion of the circle, the ray again becomes visible.

The following diagram illustrates the effect of

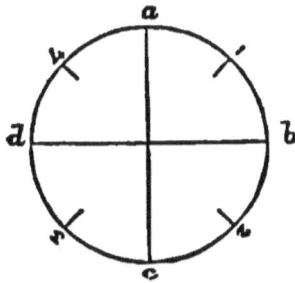

Fig. 21.

the various positions of the analyzer. At a the ray is visible, at b invisible, at c visible, at d invisible; as the analyzer passes from *a* to *b*, the brightness of the image gradually diminishes; from *b* to *c* the brightness increases. The positions marked 1, 2, 3, 4, are called the neutral axes, only half the amount of light being transmitted.

In order to analyze a polarized beam it is not necessary that the analyzer should be of the same material as the polarizer; a reflected ray may be examined by a tourmaline or crystal of Iceland spar, and a refracted or transmitted ray can be reflected from the surface of a mirror.

The student will have gathered from what we have stated in the preceding pages, that the effect of an analyzer on a polarized ray is the alternate transmission and stoppage of that ray. The most gorgeous effects are, however, obtained when a doubly refracting film is interposed between the polarized ray and the analyzer, producing what is termed "chromatic polarization."

This doubly refracting film receives the polarized ray, and doubly refracts it; in other words, the series of undulations of which the ray is composed on entering the film (sometimes called the depolarizer) is broken into two systems within it, forming the ordinary and extraordinary rays.

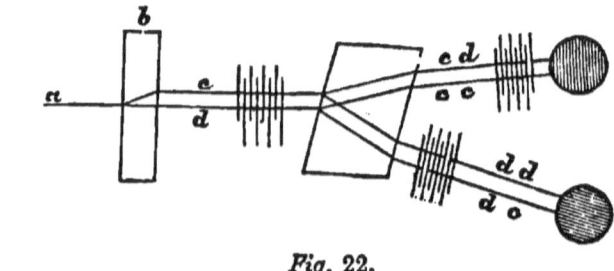

Fig. 22.

If a polarized ray is allowed to enter a film of selenite, it becomes refracted, and forms two distinct rays. *a* is a polarized ray, *b* the film of selenite, *c* is the extraordinary ray, *d* the ordinary ray; but one of these rays is retarded. If they are analyzed by a double-image prism, the ordinary and extraordinary rays will again be divided into *c d, c c* and *d d, d c;* and if the original ray be passed through a circular aperture, two coloured discs will be observed, the colour depending upon the thickness of the selenite film. If one disc is red, the other will be green, the colours being complementary to each other.

When a plate of tourmaline or a Nichol's prism is used, one of these rays is alternately suppressed. If the analyzer is revolved, we shall find that when the angles of the polarizer and analyzer coincide, and supposing a red and green selenite is used, the colours will appear in the following order:—

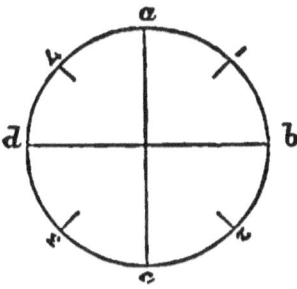

Fig. 23.

At *a* the ray would be green, at *b* red, at *c* green, at *d* red; as the analyzer approached 1, the colour fades; when it reaches that position the colour will disappear; as it approaches *b*, the red increases in brilliancy until it reaches *b*, when it will have reached its maximum brightness. In the positions 2, 3, and 4, no colour will be found.

POLARIZED LIGHT. 105

Having endeavoured to describe as plainly as possible the nature of polarized light, we will now proceed to describe the methods usually adopted for the purpose of applying polarized light to the examination of microscopic objects.

The micro-polariscope usually consists of two Nichol's prisms, mounted in appropriate fittings. A Nichol's prism is composed of a crystal of Iceland spar. It will be remembered that a beam of light, in passing through a rhomb of Iceland spar, becomes doubly refracted, and both polarized beams are visible; for polarizing purposes this is not by any means desirable, but the difficulty has been overcome in the following manner. A rhomb is divided, as shown in fig. 24, and the two halves cemented with Canada balsam; the refractive power of the film of balsam being different to that of the spar, throws the second image out of the field.

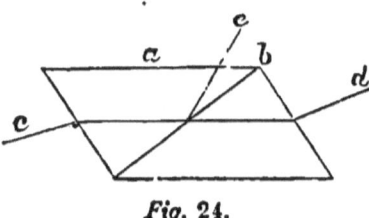

Fig. 24.

a, section of Nichol's prism; *b*, film of balsam; *c*, ray of light; *d*, ditto passing out parallel to that of incident ray; *e*, refracted ray.

a represents a section of a Nichol's prism, *b* the cementing film of Canada balsam, *c* a ray passing into the prism, *d* the same passing out parallel to the incident ray, *e* the refracted ray.

One of these prisms is mounted, as shown in fig. 25, and is made to slide in the short tube attached to the under side of the stage; *a* is a revolving collar connected with the tube into which the prism is fitted. By this contrivance

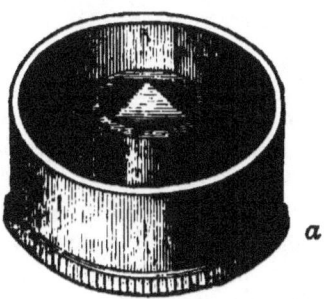

Fig. 25.

the surface of the prism can be placed at any angle with the analyzer.

The prism used as the analyzer is sometimes mounted in a brass cap, fitted over the eye-piece, as in fig. 26; or in an adapter screwed on to the

Fig. 26.

nose-piece of the Microscope into which the object-glass is screwed.

By the first method the brightness of the field and the definition of the object under examination is not impaired, but the diameter of the field is seriously diminished. By the latter plan the field remains the same size, but a certain amount of definition is sacrificed (this, however, is scarcely perceptible if the prism is of good quality). Having fixed the polarizing apparatus to the Microscope, we may now proceed to test its effects on various objects. Some will be tinted with all the colours of the spectrum, whilst others are either not affected by the altered condition of the light, or are merely black on a white ground, or white on a black ground. The last-named objects are best seen with a film of selenite placed beneath. This is sometimes mounted between two ordinary glass slides and placed below the object. The selenite should, however, be mounted in such a way that it can be revolved independently of the object. This is done in several ways; the best contrivance is perhaps the revolving selenite stage. The following diagram represents one of the simplest forms of revolving stage.

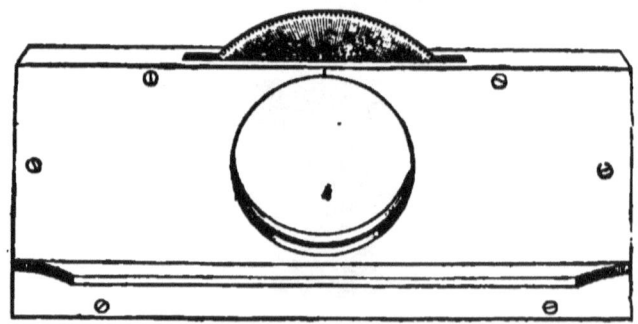

Fig. 27.

With these stages a set of selenites is usually supplied; these separately give the blue, purple, and red, with their respective complementaries orange, yellow, and green.

These discs generally have engraved upon them the amount of the retardation of the undulations of white light thus—$\frac{1}{4}$, $\frac{3}{4}$, and $\frac{9}{4}$; and if these are placed so that their positive axes (marked *P A*) coincide, they give the sum of

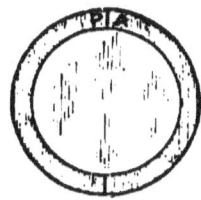

Fig. 28.

their combined retardations. If any be turned until its *P A* is at 90° to the *P A* of the others, the lesser number is subtracted from the greater. For instance, when the *P A* of the $\frac{3}{4}$ is placed at right angles to the *P A* of the $\frac{9}{4}$ the sum of the difference is obtained $= \frac{6}{4}$; if the $\frac{1}{4}$ is now added with its *P A* coinciding with the *P A* of the $\frac{9}{4}$, $\frac{7}{4}$ are obtained; but if placed to coincide with the *P A* of the $\frac{3}{4}$, $\frac{5}{4}$ is the result.

Therefore by subtracting by 90°, or adding by the *P A*, any number from $\frac{1}{4}$ to $\frac{13}{4}$, undulations may be retarded which includes all the colours of the spectrum.

To those who may wish to try the effect of polarized light at a small cost, the following plan, suggested by Professor Reinicke* will be found useful.

Procure from twenty to twenty-five pieces of thin covering glass flat and free from veins. The size most convenient for the purpose is 18 × 12 mm. Fig. 29 represents the exact size. These are to be fixed on a tube at an angle to the tube of

---

* The Professor says 50 to 60, but with that number the loss of light is considerable.

35° 25″. This tube may be made of cardboard, as shown in Fig. 30 (also the exact dimensions). The

Fig. 29.

width from a to b, and c to $d = 12$mm., that from b to c, and d to e, equal to the length of the thin glass, when placed at the proper angle. It will be found convenient to cut the cardboard partially

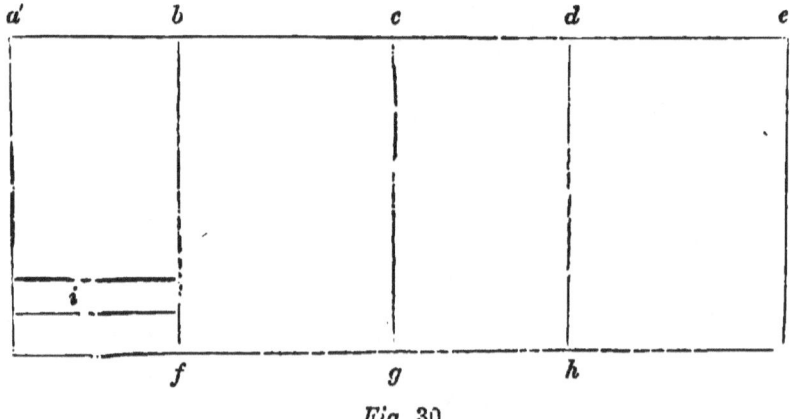

Fig. 30.

through with a sharp knife from b to f, c to g, and d to h; near the bottom of the first division on the other side paste a strip of card i; carefully paste the two edges of the card together; drop the pieces of glass into the tube, taking care that the lower edge of the first piece rests on the cardboard ledge i. When all the pieces are in position a similar strip of card must be pasted on the upper part of the opposite side of the tube. The analyzer can of course be constructed the same way. These square tubes can be fitted into cylindrical ones, and adapted to the fittings of the Microscope.

Although with this form of polariscope the young student will be able to examine many objects by polarized light, the Nichol prisms are far superior for the purpose, and most of the opticians supply the polarizing apparatus for students' Microscopes at a moderate cost (from 30s. to 35s.).

Having now described the Micro-polariscope, and the mode of using it, we will proceed to describe a few of those objects to which polarized light may be effectively applied. Matter possessing a crystalline structure as a rule affords the greatest variety of form and colour. The following list of salts, &c., most of which are easily procured, give a brilliant display of colour when polarized:—

    Chloride of Barium.*
    Chlorate of Potash.*
    Sulphate of Copper,*
     ,, ,, Nickel.*
     ,, ,, Iron.*
     ,, ,, Zinc.*
     ,, ,, Lime.
    Tartrate of Soda.*
    Salicine.
    Iodo-sulphate of Quinine.
    Asparagine.

Succinic acid.
Stearine.
Picrate of Aniline.
Chlorate of Cinchonine.
Borate of Soda.
Margarine.
Quinidine.
Santonine.
Sugar.
Uric acid.
Chromate of Potash.
Paraffine.
Platino-cyanide of Magnesium.

The beginner need not make use of a large quantity of the material he is about to experiment with, and the only apparatus he requires is a small test-tube about 4 inches long and half an inch in diameter. Fill about 1 inch of this with distilled water (if the crystals are soluble in water), add two or three crystals, and dissolve with heat if necessary; take up a small quantity with a dipping tube and drop it on a *perfectly clean* slide or cover. It is as well to prepare several slides, allowing some to dry slowly, and others to be evaporated over a spirit-lamp.

One of the most beautiful examples of crystallization is that of Salicine, and as merely recrystallizing it from its solution will only result in disappointment, we will give explicit directions for the production of the rosette form of crystals as in Fig. 2, plate 9. A saturated solution of the alkaloid must be prepared, a drop of the solution placed on a glass cover, and held over a spirit-lamp until it not only evaporates the water, but melts the residuum. The cover must now be put in a cool place, and protected from dust. If the cover is examined after the lapse of a short time, small circular, semi-transparent spots will be found scattered over the surface. Further crystallization may be prevented by warming it and mounting in Canada balsam or Dammar.

The iodo-sulphate of Quinine (Fig. 1, plate 9), also requires special preparation.

These crystals were first prepared, and their optical properties described by Dr. Herapath, of Bristol. The following are his own directions for making them :—

Mix 3 drachms of pure acetic acid with 1 drachm of alcohol ; add to these 6 drops of diluted sulphuric acid (1 to 9).

One drop of this fluid is to be placed on a glass slide, and the merest atom of quinine added, time given for solution to take place; then, upon the tip of a very fine glass rod, a very minute drop of tincture of iodine is to be added. The first effect is the production of the yellow or cinnamon-brown coloured, composed of iodine and quinine, which shows itself as a small circular spot ; while the alcohol separates in little drops, which, by a sort of repulsive movement, drive the fluid away. After a time the acid liquid again flows over the spot, and the polarizing crystals of iodo-sulphate of quinine are slowly produced without the aid of heat.

Dr. Herapath also succeeded in producing these crystals in large plates, which could be used in place of tourmalines, and they are called artificial tourmalines or Herapathite.

Santonine is an alkaloid prepared from the so-called *Semen Cynæ*, or worm seed. It is soluble in alcohol, chloroform, and water. Each solvent alters the character of crystal. With chloroform the crystals assume a lace-like appearance; crystallized from water, they arrange themselves in tufts, composed of small oblong plates, arranged round a nucleus. Santonine may also be crystallized on a hot slide, when crystals radiating from a centre will be formed.

Asparagine, an alkaloid obtained from asparagus, crystallizes in diamonds similar to the crystals of Aspartic acid, shown in fig. 3, plate 9. This acid is obtained from asparagine, but is difficult to procure; a specimen had therefore better be procured from the dealers in microscopic objects.

Succinic acid is obtained by the distillation of amber.

The preparation of slides of paraffine, stearine, margarine, and wax offer no difficulties to the beginner; all that is necessary is to place a small piece of the material on a warm slide; then place a thin cover over it, heat the slide until the substance melts, press down the cover, continuing the pressure until the slide is cold; or the slide can be placed at once on the stage of the microscope, and the gradual crystallization observed as the slide becomes cold.

The medium in which a salt is dissolved affects the form and arrangement of the crystals when it is recrystallized. The media affording the best results are gelatine, gum, and albumen.

The following method will enable the young student to add many beautiful slides to his collection of polariscope objects. Dissolve, with heat, a small piece of gelatine in the test-tube before described, using a similar quantity of distilled water. In another test-tube make a saturated solution of the salt (sulphate of copper, for example), add a few drops to the gelatine (mix thoroughly, but avoid forming bubbles, stirring it with a glass rod or piece of platinum wire); spread a drop on a glass cover, set aside in a cool place to dry; this will usually take about half an hour. If the experiment has been successful, the crystals will appear like fern fronds. (See fig. 4, plate 9.)

This figure will give some idea of the elegance of form and beauty of colour; but it is beyond the skill of any artist to do justice to the beauty of a good slide. The sulphates of nickel and iron are also very good when crystallized out of gelatine.

With chlorate of potash, a totally different form of crystallization is produced, the crystals being tabular and large. A very remarkable effect is produced when a small quantity of a solution of barium is added; the barium will be found to have crystallized in small moss-like tufts at the angles. Chloride of barium mixed with the gelatine solution assumes a dendritic form, somewhat resembling sulphate of copper, but polarizes differently. Gum arabic may be substituted for gelatine; the *modus operandi* is, however, similar; albumen (white of egg) requires to be dried before it is added to the distilled water, which must be only slightly warmed. The student cannot do better than try the effect of the different media; some salts do better with gelatine, others with gum; for example, he will be able to produce more effective slides of tartrate of soda with gum than gelatine.

Platino-cyanide of Magnesium must be prepared without heat, as warmth alters the colour of the crystals. We have obtained the best results by adding a few crystals to a drop of the gelatine solution previously placed on the slide, stirring them with a stout bristle until dissolved, and then allowing them to slowly recrystallize. These crystals, like the iodo-sulphate of Quinine, will analyze a polarized ray. They are best mounted in dammar.

Very beautiful results may be obtained by a mixture of two or more salts. Mr. Davies, in the *Quarterly Microscopical Journal*, Vol. II., N.S., gives the following directions for crystallizing the double sulphate of copper and magnesia. Make nearly a saturated solution of the two salts, place a drop on a slide and dry rapidly, allowing the slide to become hot enough to fuse the salt, which will now appear as an amorphous film on the slide. On slowly cooling, the salt will absorb moisture from the surrounding air, and crystallization will commence from various points, assuming the appearance of flowers. As soon as these "flowers" are perfected the slide should be slightly warmed and a little of *pure* Canada balsam dropped upon it, and covered with the usual thin glass cover.

Sugar requires a somewhat different treatment to any of the crystals previously described, and the tyro's first attempts will probably result in disappointment. The best for the purpose is the white "stone sugar." Dissolve this in water, using enough to form a thick syrup; spread a drop on a cover, drying it quietly over a spirit-lamp; when dry place it in a damp cellar or cupboard. In the course of twenty-four hours crystallization will have taken place. The cover should now be mounted in balsam.

Passing from the inorganic to organic we proceed to give a few hints on the preparation of specimens from vegetable and animal kingdoms.

The following are a few of the objects from the former which the student will have little difficulty in obtaining :—

        Potato starch.
        Tous les mois ditto.

Cotton fibre.
Hairs and scales from leaves.
Longitudinal sections of wood.

The first-named on our list can be very easily procured by scraping a potato, and then shaking the pulp in a test-tube with water, to which a small quantity of soda has been added. The starch will rapidly subside, and the fibrous matter, &c., can be poured off. The washing should be repeated until the starch is left perfectly pure. Starch for polarizing purposes requires to be mounted in Canada balsam. The hairs and scales of plants require no preparation for mounting. The scales of *Eleagnus* or *Hippophæ rhamnoides* (sea buckthorn) (Fig. 6, pl. 9), are easily procured, and offer no difficulty to the young manipulator, merely requiring to be detached from the cuticle of the leaf with the point of a knife or lancet, and afterwards transferred to a drop of water, to which a minute quantity of gum has been previously added, when dry, mount in balsam.

The following list contains a few of the objects from the animal kingdom :—

Fish scales.
Palates of Mollusca.
Hairs.
Quill.
Horn.
Whalebone.

Fish scales are so well known that no difficulty can arise in obtaining specimens, with the exception of those from the eel, which do not occur on the surface, but will be found imbedded in the skin; they may be obtained by picking the skin with the point of a needle, previously scraping off the mucus.

The palates of mollusca, as polariscope objects, are not as a rule very effective, that of the common Whelk excepted.

Hairs are worthy of notice for polarizing purposes, as they usually display a considerable amount of colour. While horsehair and grey human hair (Fig. 5, pl. 9) are perhaps the best for the student's purpose.

The structure of Rhinoceros horn and whalebone is well displayed when polarized, but they are difficult to prepare, and it would be better to purchase them of the dealers in microscopic preparations.

The mineral kingdom affords but few objects that can be prepared by the amateur, although no collection of polariscope objects would be complete without one or more sections of agate and chalcedony; and, like the objects previously named, must be obtained from the dealer. The young microscopist should, however, obtain from some optician a piece of the so-called Brazilian pebble (really transparent quartz), and break it up into small fragments: many of these, when mounted, display very beautiful coloured rings.

A few words may perhaps be necessary as to the mode of procedure in mounting specimens of crystal. We have in several instances directed the solution to be placed on the cover; our reason for doing so is, in order to avoid the application of any great degree of heat, and at the same time using tolerably hard balsam.

The plan we adopt is as follows. Place a drop of pure balsam on the centre of a slide, harden over the lamp (it will be sufficiently so if the nail slightly indents it when cold); now drop a little turpentine on the prepared cover, holding it as close as possible to the edge with the forceps, rewarm the slide, and apply the opposite edge of the cover to the edge of the balsam, and allow the cover to fall gradually down; when the glass disc is covered by the balsam, press carefully until all the superfluous balsam is squeezed out.

We must now, however, draw our last half-hour to a close. All we have attempted has been in the way of introduction. We have only described those things which are most easily obtained, and we have sought rather to create a desire for further knowledge, than to impart an exhaustive amount of information on any one subject.

Those who have properly apprehended our remarks will see that there is not a distinct science of microscopic objects, but that these objects belong to various departments of science, whose great facts and principles must be studied from works devoted to them. The Microscope is in fact an instrument to assist the eyes in the investigation of the facts of structure and function, wherever they may occur in the great field of nature; and that inquirer must have a very limited view of the nature of science, who supposes either that the Microscope is the only instrument of research, or that any investigation, where its aid reveals new facts, can be successfully carried on without it.

APPENDIX.

BY THOMAS KETTERINGHAM.

THE PREPARATION AND MOUNTING OF OBJECTS.

THE majority of objects exhibited by the Microscope require some kind of preparation before they can be satisfactorily shown, or their form and structure properly made out. To convince the beginner of this, let him take the leg of any insect, and, without previous preparation, place it under his Microscope, and what does he see? A dark opaque body, fringed with hair, and exceedingly indistinct. But let him view the same object prepared and permanently mounted, and he will then regard it with delight. That beautiful limb, rendered transparent by the process it has undergone, now lies before him, rich in colour, wonderful in the delicate articulation of its joints, exquisite in its finish, armed at its extremities with two sharp claws equally serviceable for progression or aggression, and furnished, in many instances, with pads (*pulvilli*) (see plate 7, figures 205, 206), which enable the insect to walk with ease and safety on the smoothest surface. If the beginner has a true love for the study of the Microscope, he will be glad of information respecting the method pursued in dissecting and preserving microscopic objects, nor will he rest satisfied until he has acquired some knowledge of the art. We will briefly point out a few of the advantages possessed by those who are able to prepare specimens for themselves.

Objects well mounted will remain uninjured for years, and will continue to retain their colour and structure in all their original freshness.

They can be exhibited at all times to one's friends, and may be studied with advantage whenever an opportunity occurs.

By the practice of dissection such a knowledge is gained of the varied forms and internal organization of minute creatures as can be obtained in no other way.

There are doubtless many who, possessing a small Microscope, are unable by reason of their limited means to expend money in the purchase of ready-prepared specimens. To such, a few plain directions, if followed, will be of service, and will enable them to prepare their own.

The materials necessary for the beginner are few, and not expensive. In fact, the fewer the better; for a multiplicity is apt only to cause confusion. The following will be found sufficient for all ordinary purposes, and may be obtained at any optician's.

Bottle of new Canada balsam.
Bottle of gold-size.
Bottle of Brunswick black.
Spirits of turpentine—small quantity.
Spirits of wine—small quantity.
Solution of caustic potash (*liquor potassæ*).
Ether—a small bottle.
Empty pomatum-pots, with covers, for holding objects while in pickle.
Half a dozen needles mounted in handles of camel-hair brushes.
Pair of brass forceps.
Two small scalpels.
Pair of fine-pointed scissors.
Camel-hair pencils—half a dozen.
Slips of plate-glass, one inch by three inches—two dozen.
Thin glass covers, cut into squares and circles—half an ounce.

We will suppose that the beginner, having purchased the necessary materials, is about to make his first attempt. Let him attend to the following advice, and he will escape many failures.

He must bring to his work a mind cool and collected; hands clean and free from grease. Let him place everything he may require close at hand, or within his reach. A stock of clean slides and covers must always be ready for use. He must keep his needles, scissors, and scalpels scrupulously clean. An ingenious youth will readily construct for himself a box to contain all his tools. Cleanliness is so essential to success, that too much stress cannot be laid upon it. All fluids should be filtered and kept in well-corked phials. A bell-glass, which may be purchased for a few pence, will be found exceedingly useful in covering an object when any delay takes place in the mounting. For want of it, many specimens have been spoilt by the intrusion of particles of dust, soot, and other foreign substances. Let the table on which the operator is at work be steady, and placed in a good light, and, if possible, in a room free from intrusion.

WINGS OF INSECTS.—Perhaps these are the easiest objects upon which the beginner can try his "'prentice hand." Here little skill is required. Select a bee, or wasp, and with your fine scissors sever the wing from its body; wash it with a camel-hair brush in some warm water, and place it between two slips of glass, previously cleaned, which may be pressed together by a letter-clip, or an American clothes-peg; place it in a warm corner for a few days; when *quite dry*, remove it from between the slides, and soak it for a short time in spirits of turpentine. This fluid renders the object more transparent, frees it from air-bubbles, and prepares the way for a readier access of the balsam to the various portions of its structure.

Having selected from your stock a clean slide of the requisite size, and a thin glass cover somewhat larger than the object about to be mounted, hold them both up to the light, when any slight impurities will appear, and may be speedily removed by rubbing the surfaces of the glass with a fine cambric handkerchief, or a piece of soft wash-leather. Should, however, a speck or flaw in the glass itself be found in the centre of the slide, at once reject it and choose another. Remove the wing with a pair of forceps from the turpentine, and place it in the exact centre of the slide: this may be accomplished by cutting a stiff piece of cardboard, tin, or zinc, the size of the slide, and punching a hole, the edge of which should be equally distant from each end and each side; lay the slide upon it, and place the object in the circular space; you will thus get it properly centred.

Before dropping the balsam (which should have been previously warmed) upon the specimen, place it under the Microscope: you may possibly detect some foreign substance, in the shape of a particle of soot or a fibre from your handkerchief, in contact with it; remove it with the point of a needle. Take up a small quantity of the balsam on the end of a small glass rod, and let it fall upon the object; hold the slide for a few minutes over the flame of a candle or spirit-lamp at a distance sufficient to make it warm, but not hot; the balsam will gradually spread itself over and around the object: should air-bubbles arise, they may be broken by touching them with the point of a needle; they will, howver, frequently disperse of themselves as the balsam dries. The thin glass cover, being warmed, should now be placed upon the object, and a slight pressure applied to get rid of the superfluous balsam. Place the slide in some warm spot to dry; an oven will do very well, if the fire has been some time removed and there is not sufficient heat to make the balsam boil.

In a short time the balsam round the edges of the cover will be hard enough to admit of the greater part being scraped off with a knife; the remainder may be got rid of by wiping the slide with a rag dipped in turpentine or ether. The finishing touch consists in labelling the object with its proper name. It will be found advantageous to place the common name of the specimen at one end of the slide, and its scientific name at the other.

Some persons prefer covering their slides with ornamental paper, which may be obtained of almost any optician. Others prefer the glass without any covering at all. In the latter case the edges of the slide should be ground, the round thin glass covers used, and the name scratched upon the slide with a writing diamond. In the former, the edges of the slide, being covered with paper, need not be ground, but square thin covers should be used instead of round ones, and the name written with pen and ink in the square places allotted at each end of the slide.

LEGS OF INSECTS (plate 7, figures 205, 206, 207; plate 8, figures 215 to 219, 223, 224.—These require a little more preparation than wings; and as they possess some thickness, and are mostly opaque, besides being of a hard, horny character, they should be placed for a fortnight, or even longer, in *liquor potassæ;* this will soften the tissue and dissolve the muscles and other matter contained within them, so that by gently pressing the limb between two slips of glass, the interior substance will gradually escape, and may be removed by repeated washings. The squeezing process, however, must be conducted gently, to prevent any rupture: perhaps the best plan is to plunge the slips of glass into a basin of clean water, when all impurities oozing out from the pressure will sink to the bottom. Should the leg not be sufficiently softened to be squeezed quite flat, it must be again placed in the solution for a longer period, until this result be obtained. On removing it from the potash, it should be well washed with a camel-hair pencil in clean water, placed between two slips, held together by an American clothes-peg with a good stiff spring. If placed in a warm corner, a few days will be sufficient to dry it thoroughly: afterwards soak it in spirits of turpentine; the time of immersion to be regulated by the opacity of the object.

The directions for mounting in balsam are precisely the same as those given for the wings of insects. Care should be taken not to heat the balsam too hot, as it will invariably destroy delicate specimens by curling them up. In tough horny structures, such as the wing-cases of beetles, &c., heat is sometimes an advantage, and there are a few structures that show to advantage when the balsam has been heated to a boiling pitch; but for the majority of objects, a gentle warmth is all that is required.

OVIPOSITORS AND STINGS (plate 7, figure 200) are more difficult to prepare, and require some amount of dissection before they can be properly displayed. To do this, some degree of skill is necessary, and a knowledge of insect anatomy, which can be acquired only by study and practice. As a rule, all dissections should be carried on as far as possible with the naked eye; when this has been accomplished, we must then seek the aid of lenses.

The object-glasses of one's Microscope are the best that can be used for the purpose. An inch lens will be found especially fitted for the work. A simple Microscope, provided with a broad stage, and an arm movable by rack and pinion, for carrying the lenses, is the kind of instrument usually employed. It should be strongly made, and capable of bearing a good deal of rough usage.

Dissections may be carried on under the compound Microscope; but we do not think the beginner would succeed, as objects become inverted and motion reversed when seen through this instrument. If, however, it be provided with an erector, this difficulty is overcome by the object being brought into the same position that it occupies when seen by the naked eye.

As most dissections are carried on under water, some kind of shallow trough is necessary to contain it: watch-glasses answer the purpose remarkably well. The small white dishes and covers used for rubbing up colours will be found very useful; also some cork bungs on which to pin the object; and these last should have their under sides loaded with lead to sink them in fluid. A great many delicate dissections may, however, be made in a drop of water placed on a slip of glass; but for all objects of large size, the trough, or some similar contrivance, will be necessary.

All insects that have been killed a long time, and whose bodies are hard and brittle, may be softened by immersing them in the solution already mentioned.

The sting of the bee, wasp, hornet, and the ovipositors of many flies, especially the ichneumons, are very similar in their structure, and are generally found at the termination of the abdomen, from which they may be obtained by first slitting open the body of the insect with the fine scissors, and afterwards removing the sting by using the scalpel and needles. One or two of the latter should have their points curved, which may easily be accomplished by heating the ends red-hot in the flame of a candle, and bending them with a pair of small pincers. At first sight the sting presents nothing to the eye but a horny sheath, tapering to a point, with a slit broadest at its base and running down the entire length; within this sheath, on each side, lies a barbed, sharp-pointed spear, in large insects capable of inflicting a severe wound, while the tube in which they are lodged acts as a steadying rod, and as a channel to conduct a virulent poison to the wound. The bag containing the poison is placed at the root of the sting, and is connected by a narrow neck with the sheath. The difficulty in the dissection of the sting lies in getting the barbed points out of the sheath and placing them on each side of it. The following is the method employed by the writer. The sting is placed in potash until it loses some of its rigidity; it is then transferred to a slip of glass or earthenware trough. The curved needle-points are essential here. With one, hold the object firmly on the stage of the Microscope, insert the point of the other into the opening at the base of the sheath where it is largest, and gradually draw the point down the tube; this will make the opening wider, and dislodge the barbs; arrange them on each side of the sheath, place the sting between two glass slips subject to pressure. When dry, soak it for a few days in turpentine, and mount in balsam in the usual manner. A good specimen ought to show the barbs very distinctly on each side of the sheath.

It will be found useful to the student to prepare three specimens of this organ :—

1st. The whole abdomen, showing the position the sting occupies within it.

2nd. The sting with the barbs lying within the sheath.

3rd. The barbs pulled out of the sheath and placed on each side of it.

Three such specimens well mounted will enable the student

to study the structure of this curious organ with advantage.

SPIRACLES (plate 7, figures 212, 213).—These do not require much dissection. They are generally found on each side of the abdomen, almost every segment of which possesses a pair. Excellent specimens are furnished by the dytiscus, bee, blowfly, cockchafer, and silkworm. To prepare them, separate from the thorax the abdominal portion of the insect, and slit it down the centre with the fine-pointed scissors, draw out the viscera, &c., with the curved needles. The air-tubes adhering to the spiracles may be detached by cutting them away with the scissors. Thoroughly cleanse the horny cuticle by repeated washings, spread it out flat between two slips of glass; when dry, immerse it in spirits of wine or turpentine for a few days, and mount it in balsam. In this manner the whole of the spiracles of an insect, running down each side of the abdomen, will be displayed.

TRACHEÆ (plate 8, figure 222).—The best method we are acquainted with for obtaining the air-tubes of insects is that recommended by Professor Quekett:—

"By far the most simple method of procuring a perfect system of tracheal tubes from the larva of an insect, is to make a small opening in its body, and then to place it in strong acetic acid: this will soften or decompose all the viscera, and the tracheæ may then be well washed with the syringe, and removed from the body with the greatest facility, by cutting away the connections of the main tubes with the spiracles by means of the fine-pointed scissors. In order to get them upon the slide, this must be put into the fluid and the tracheæ floated upon it, after which they may be laid out in their proper position, then dried, and mounted in balsam."

The best specimens are found in the larva of the dytiscus and cockchafer, and in the blowfly, goat-moth, silkworm, and house-cricket.

GIZZARDS (plate 8, figures 220, *a*, *b*; 221, *a*, *b*).—Most of the insects from which these organs are procured being of large size, it will be necessary to secure them to one of the loaded corks by small pins. The dissection should be made in one of the shallow troughs, filled with weak spirits and water. Cut the insect open; the stomach will float out with the gizzard attached to it, in the shape of a small bulbous expansion of the size of a pea. Insert the fine point of the scissors, and cut it open; the interior will be found full of food in process of trituration. Empty the contents of the gizzard, and wash it out well; place it for a few days in the solution of potash: and, finally, cleanse it with some warm water and a camel-hair brush. Spread it out flat between slips of glass; when dry, place it in turpentine for a week, and afterwards mount it in balsam.

The best specimens for displaying the horny teeth with which the gizzard is furnished are obtained from crickets, grasshoppers, and cockroaches.

PALATES (plate 6, figures 171, 172, 173, 174).—These consist of a narrow kind of tongue, armed with a series of horny teeth, placed in regular rows. The whelk, limpet, periwinkle, garden-snail, and the snails found in our cellars and aquariums, are all furnished with this peculiar apparatus, which may be obtained by laying open the body with the scalpel or scissors. It will generally be found curled up near the head, and may be distinguished by its ribbon-like appearance: patience and skill are necessary to extract it from the surrounding mass. When properly cleaned, it may be at once pressed flat and dried between slips of glass. Many palates polarize well when mounted in balsam; but if not intended for polarization, they should be mounted in a preservative fluid, composed of five grains of salt to one ounce of water.

TONGUES, PROBOSCES, MANDIBLES, AND ANTENNÆ (plate 7, figures 197, 199, 202 to 204) are amongst the most beautiful objects exhibited by the Microscope. Many of these, besides the ligula, possess several sharp lancets for puncturing the skin of animals from whom they derive their sustenance. To arrange these organs so that each part may be clearly seen, requires a good deal of delicate manipulation. It is generally more satisfactory to mount the whole head of the insect. To accomplish this, it must be softened by immersion in *liquor potassæ* for some time, and the interior substance got rid of by pressure. To dry it flat, place it between two slips of glass, which should be held together by a spring-clip; soak it for a fortnight or longer in turpentine, until it becomes transparent, and then mount it in balsam.

The head of the bee, wasp, dronefly, blowfly, and gadfly, are all excellent examples of the varied structures of these suctorial organs.

EYES (plate 7, figures 208, 208a, 210).—The compound eyes of insects, for the display of their numerous facets, should be dissected from the head, and macerated in fluid. The black pigment lining the interior may be got rid of by washing it away with a camel-hair brush. When quite clean, the cornea may be dried and flattened between two slips of glass. In practice, however, the cornea, from its sphericity, will be found to have a tendency to fold in plaits, or to split in halves. To remedy this, cut with the fine scissors a few notches round its edges; it may then be flattened without danger of its either wrinkling or splitting. When the cornea is very transparent it should be mounted in a cell with some kind of preservative fluid (spirit and water will do very well), otherwise the structure will be lost if mounted in balsam, the tendency of that substance being to add transparency to every object with which it comes in contact. But there are many insects in whose eyes the hexagonal facets are strongly marked: all such will show best when mounted in balsam.

HAIRS (plate 7, figures 184 to 191).—These may be mounted either in fluid or balsam, first taking the precaution to cleanse them from fatty matter by placing them in ether. If the hair be coarse and opaque, mount it in balsam; if fine and transparent, it should be mounted in a cell, with some weak spirit.

Sections of hair are made by gluing hairs into a bundle, and placing it in a machine for making sections. By means of a sharp knife which traverses the surface, the thinnest slices may be cut, and each individual section afterwards can be separated in fluid. To select the thinnest and best, place them under the Microscope. The point of a camel-hair pencil will be found the best instrument for transferring them to a clean slide. When dry, mount them in balsam, as usual. Some very good sections of the hairs of the beard may be obtained by passing the razor over the face a few minutes after having shaved.

SCALES OF FISH (plate 6, figures 178 to 180).—These dermal appendages may be detached from the skin by a knife; and if to be viewed as opaque objects, may be dried and mounted with no other preparation than cementing over them a thin glass cover. If intended to be viewed as transparent objects, the scales should be properly cleaned, dried, and mounted in balsam; but the most satisfactory way of exhibiting their structure is to mount them in a cell with some preservative fluid.

SCALES OF BUTTERFLIES, MOTHS, &c. (plate 8, figures 225 to 229).—Select the wing of a living or recently-killed insect, gently press it on the centre of a clean glass slide. On removing the wing, numerous scales will be seen adhering to the slide; place over them one of the thin glass covers, and cement it down by tipping lightly the edges with gold size.

Specimens should be taken from various parts of the wings of the same insect, as the form of the scales vary according to the position they occupy in the wing.

SECTIONS OF BONE (plate 8, figure 232).—All hard and brittle substances from which thin slices cannot be made by a sharp knife, must be reduced to a transparent thinness by the process of grinding down. Having selected the bone from which the section is about to be made, a thin slice should be cut from it with a fine saw. At first the section may be held by the fingers while grinding down one of its surfaces on a coarse stone; but when it approaches the thinness of a shilling, it must be cemented by some old and tough Canada balsam to a slip of glass. Upon the perfect adhesion of the section to the slide depends in a great measure the success of the operation. Having reduced the thickness of the section by a coarse stone or a file, transfer it to a hone; a few turns will obliterate scratches, and produce an even, smooth surface, which may be further polished by rubbing it on a buff-leather strop charged with putty-powder and water. When dry, attach the polished surface to the glass slip: this gives a firm hold of the section, which would otherwise become too thin to be held by the fingers. In rubbing down the unfinished surface, take care that an equal thickness prevails throughout the section. As it approaches completion, recourse must be frequently had to the Microscope, in order to determine how much further it is necessary to proceed, a few turns either way at this stage being sufficient to make or mar the specimen. When it has become so transparent that objects may be readily seen through it, remove it from the hone and polish it on the strop. To detach it from the slide when finished, place it in turpentine or ether, both being excellent solvents of balsam. Mount in the dry method, by simply cementing a thin glass cover over it. In recent bone, this method of mounting, though the most difficult, is decidedly the best for displaying its structure. Fossil bone, however, where the interstices are filled with earthy matter, shows best in balsam.

SPINES OF THE ECHINUS (plate 5, figures 151, 152); SECTIONS OF SHELL (plate 6, figures 165 to 169).—These are cut and reduced in the same manner as sections of bone; but they require greater care in grinding, in consequence of being more brittle. The polishing, however, may be dispensed with, and the section mounted in balsam.

STONES OF VARIOUS KINDS OF FRUITS (plate 8, figure 243) will well repay the labour bestowed in producing good sections. The saw, the file, and the hone are the principal agents used in the reduction of these hard osseous-like tissues. A perfect section should have but one layer of cells, which may be admirably seen when mounted in a cell with weak spirit.

SECTIONS OF WOOD (plate 3, figures 54 to 59).—To make thin sections of hard wood it will be necessary to employ some kind of cutting machine. There are several of these, more or less expensive, but the principle of construction in all is similar. The wood, after some preparation, and being cut to the requisite length, is driven by a mallet into a brass cylinder, at the bottom of which works a fine screw with a milled head. The wood is pushed to the surface of the tube, and to any degree above it by the revolution of the screw; when a sharp knife, ground flat on one side, is brought with a sliding motion in contact with it. The slices may be removed from the knife by a wetted camel-hair pencil, placed in some weak spirit, and examined at leisure; the thinnest and most perfect section being retained for mounting. Green wood previous to being cut should be placed in alchohol and afterwards in water. Hard and dry wood may be made sufficiently soft for slicing by first immersing it in water for some days. Sections of the above may be mounted either in balsam or fluids. Stems of plants, horny tissues, and many other substances not sufficiently hard to be ground down, may be cut into slices of extreme thinness by this handy instrument. In order to obtain a correct idea of the structure of wood, bone, and shell, sections should be made in vertical, transverse, and oblique directions.

CUTICLE OF PLANTS (plate 2, figures 42 to 46), HAIRS (plate 3, figures 74 to 88), AND SPIRAL VESSELS (plate 2, figures 47 to 49), may all be obtained by macerating the leaves and stems of plants in water, and afterwards dissecting them with the needles. Good specimens of the cuticle, showing the stomata, may be often obtained by simply peeling off the skin with a sharp knife. Hairs may be detached from various parts of a plant by a similar process. Spiral vessels will, however, require to be separated by the needles from the surrounding tissues. All delicate vegetable preparations are best displayed when mounted in a cell with weak spirit.

Cells for mounting objects in fluid are generally formed of some kind of varnish upon which the fluid will not act; gold-size and Brunswick black are most commonly used. To form a cell, simply charge a camel-hair brush with the varnish, and enclose with a broad black ring a small circular space on the centre of the slide. When quite dry, it is ready

K

for use. Place the object, with a small quantity of fluid, in the cell; and having lightly touched the edges of the thin glass cover with gold-size, drop it gently on the specimen; the superfluous fluid will escape over the sides of the cell, and may be removed by small pieces of blotting-paper, taking care, however, that none of the fluid is drawn from the interior of the cell; in which case an air-bubble would immediately appear. To make the cell air-tight, gradually fill up the angle formed by the edges of the cover with the cell, by running several rims of varnish round it. In order to prevent the cement from running into the cell and spoiling the specimen, each layer should be dry before another is placed upon it.

The student should always have a stock of cells on hand ready for immediate use. Dozens of these cells may be made in half an hour by an ingenious little turntable, the invention of Mr. Shadbolt, and which may be obtained for a few shillings.

The limits of this little work have precluded us from giving little more than general directions respecting the permanent preparation of microscopic objects. Our object has been merely to give a few plain instructions, which, if carefully followed, will enable the beginner to prepare some of the most popular objects exhibited by the Microscope.

THE END.

www.ingramcontent.com/pod-product-compliance
Lightning Source LLC
Chambersburg PA
CBHW030256170426
43202CB00009B/763